HVAC
Pocket Reference

James Brumbaugh

Wiley Publishing, Inc.

Vice President and Executive Group Publisher: Richard Swadley
Vice President and Publisher: Joseph B. Wikert
Executive Editor: Carol A. Long
Production Editor: Angela Smith
Text Design & Composition: TechBooks

Published by Wiley Publishing, Inc., Indianapolis, Indiana
Published simultaneously in Canada

Library of Congress Control Number:

ISBN 13 : 978-0-7645-8810-5

SKY10082927_082724

Contents

Acknowledgments

The author would like to thank the many individuals and organizations for their contributions to the writing of the *HVAC Pocket Reference.* Among the corporations, associations, and government bodies supplying information and illustrations are:

ASHRAE
Bodine
BW Beckett Corporation
Carlin Combustion Technology, Inc.
Carrier Corporation
Copper & Brass Research Association
Copper Development Association, Inc.
Delavan Spray Technologies
Dunham-Bush, Inc.
General Electric
Hayes-Albion Corporation
Honeywell Tradeline Controls
ITT General Controls
Janitrol
Lennox Industries, Inc.
National Oil Fuel Institute
National Propane Gas Association
Midco International Inc.
Office of Energy Efficiency and Renewable Energy/ U.S. Department of Energy

Penn Ventilator Company, Inc.
Riello Burners North America
Sarco Company, Inc.
Sundstrand Hydraulics
Taco
ThermoPride
The Trane Company
Vanguard Piping Systems, Inc.
Watts Radiant, Inc.
Webster Electrical Company, Inc.
Westinghouse Electric Coporation
Wm. Powell Company
Yukon/Eagle

Introduction

The *HVAC Pocket Reference* is designed as a quick reference guide for HVAC technicians at the job site. Every possible precaution was taken to ensure the accuracy of the technical data. All specifications and tables provided by contributing organizations were reproduced in the *HVAC Pocket Reference* without change.

It is assumed that those using the *HVAC Pocket Reference* are trained and experienced HVAC technicians or have equivalent training and experience. Above all, they should have a thorough knowledge of the safety practices required when installing, repairing, or maintaining HVAC equipment. Failure to follows safety procedures required by the manufacturers of HVAC equipment can result in damage to the equipment and loss of the manufacturer's warranty. There is also the very real possibility of serious injury or even death because the technician may be working at times with high voltage electricity or volatile fuels.

Introduction

CONVERSION TABLES

Metric and English Equivalent Measures

Measures of Length

Metric		*English*
1 meter	=	39.37 inches, or 3.28083 feet, or 1.09361 yards
0.3048 meter	=	1 foot
1 centimeter	=	0.3937 inch
2.54 centimeters	=	1 inch
1 millimeter	=	0.03937 inch
25.4 millimeters	=	1 inch
1 kilometer	=	1093.61 yards, or 0.62137 mile

Measures of Weight

Metric		*English*
1 gram	=	15.432 grains
0.0648 gram	=	1 grain
28.35 grams	=	1 ounce avoirdupois
1 kilogram	=	2.2046 pounds
0.4536 kilogram	=	1 pound
1 metric ton 1000 kilograms	=	0.9842 ton of 2240 pounds 9.68 cwt. 2204.6 pounds
1.016 metric tons 1016 kilograms	=	1 ton of 2240 pounds

Measures of Capacity

Metric		*English*
1 liter (= 1 cubic decimeter)	=	61.023 cubic inches 0.03531 cubic foot 0.2642 gallons (American) 2.202 pounds of water at 62°F
28.317 liters	=	1 cubic foot
3.785 liters	=	1 gallon (American)
4.543 liters	=	1 gallon (Imperial)

English Conversion Table

Length			
Inches	×	0.0833	= feet
Inches	×	0.02778	= yards
Inches	×	0.00001578	= miles
Feet	×	0.3333	= yards
Feet	×	0.0001894	= miles
Yards	×	36.00	= inches
Yards	×	3.00	= feet
Yards	×	0.0005681	= miles
Miles	×	63360.00	= inches
Miles	×	5280.00	= feet
Miles	×	1760.00	= yards
Circumference of circle	×	0.3188	= diameter
Diameter of circle	×	3.1416	= circumference
Area			
Square inches	×	0.00694	= square feet
Square inches	×	0.0007716	= square yards
Square feet	×	144.00	= square inches
Square feet	×	0.11111	= square yards
Square yards	×	1296.00	= square inches
Square yards	×	9.00	= square feet
Dia. of circle squared	×	0.7854	= area
Dia. of sphere squared	×	3.1416	= surface
Volume			
Cubic inches	×	0.0005787	= cubic feet
Cubic inches	×	0.00002143	= cubic yards
Cubic inches	×	0.004329	= U.S. gallons
Cubic feet	×	1728.00	= cubic inches
Cubic feet	×	0.03704	= cubic yards
Cubic feet	×	7.4805	= U.S. gallons
Cubic yards	×	46656.00	= cubic inches
Cubic yards	×	27.00	= cubic feet
Diameter of sphere cubed	×	0.5236	= volume
Weight			
Grains (avoirdupois)	×	0.002286	= ounces
Ounces (avoirdupois)	×	0.0625	= pounds
Ounces (avoirdupois)	×	0.00003125	= tons
Pounds (avoirdupois)	×	16.00	= ounces

English Conversion Table *(continued)*

Weight			
Pounds (avoirdupois)	×	0.01	= hundredweight
Pounds (avoirdupois)	×	0.0005	= tons
Tons (avoirdupois)	×	32000.00	= ounces
Tons (avoirdupois)	×	2000.00	= pounds
Energy			
Horsepower	×	33000	= foot-pounds per minute
British thermal units	×	778.26	= foot-pounds
Ton of refrigeration	×	200	= British thermal units per minute
Pressure			
Pounds per square inch	×	2.31	= feet of water (60°F)
Feet of water (60°F)	×	0.433	= pounds per square inch
Inches of water (60°F)	×	0.0361	= pounds per square inch
Pounds per square inch	×	27.70	= inches of water (60°F)
Inches of mercury (60°F)	×	0.490	= pounds per square inch
Power			
Horsepower	×	746	= watts
Watts	×	0.001341	= horsepower
Horsepower	×	42.4	= British thermal units per minute
Water Factors (at point of greatest density—39.2°F)			
Miners inch (of water)	×	8.976	= U.S. gallons per minute
Cubic inches (of water)	×	0.57798	= ounces
Cubic inches (of water)	×	0.036124	= pounds
Cubic inches (of water)	×	0.004329	= U.S. gallons
Cubic inches (of water)	×	0.003607	= English gallons
Cubic feet (of water)	×	62.425	= pounds
Cubic feet (of water)	×	0.03121	= tons
Cubic feet (of water)	×	7.4805	= U.S. gallons
Cubic inches (of water)	×	6.232	= English gallons

(continued)

English Conversion Table *(continued)*

Water Factors (at point of greatest density—39.2°F)			
Cubic foot of ice	×	57.2	= pounds
Ounces (of water)	×	1.73	= cubic inches
Pounds (of water)	×	26.68	= cubic inches
Pounds (of water)	×	0.01602	= cubic feet
Pounds (of water)	×	0.1198	= U.S. gallons
Pounds (of water)	×	0.0998	= English gallons
Tons (of water)	×	32.04	= cubic feet
Tons (of water)	×	239.6	= U.S. gallons
Tons (of water)	×	199.6	= English gallons
U.S. gallons	×	231.00	= cubic inches
U.S. gallons	×	0.13368	= cubic feet
U.S. gallons	×	8.345	= pounds
U.S. gallons	×	0.8327	= English gallons
U.S. gallons	×	3.785	= liters
English gallons (Imperial)	×	227.41	= cubic inches
English gallons (Imperial)	×	0.1605	= cubic feet
English gallons (Imperial)	×	10.02	= pounds
English gallons (Imperial)	×	1.201	= U.S. gallons
English gallons (Imperial)	×	4.546	= liters

Metric Conversion Table

Length			
Millimeters	×	0.03937	= inches
Millimeters	÷	25.4	= inches
Centimeters	×	0.3937	= inches
Centimeters	÷	2.54	= inches
Meters	×	39.37	= inches (Act. Cong.)
Meters	×	3.281	= feet
Meters	×	1.0936	= yards
Kilometers	×	0.6214	= miles
Kilometers	÷	1.6093	= miles
Kilometers	×	3280.8	= feet

Metric Conversion Table *(continued)*

Area			
Square millimeters	×	0.00155	= square inches
Square millimeters	÷	645.2	= square inches
Square centimeters	×	0.155	= square inches
Square centimeters	÷	6.452	= square inches
Square meters	×	10.764	= square inches
Square kilometers	×	247.1	= acres
Hectares	×	2.471	= acres
Volume			
Cubic centimeters	÷	16.387	= cubic inches
Cubic centimeters	÷	3.69	= fluid drams (U.S.P.)
Cubic centimeters	÷	29.57	= fluid ounces (U.S.P.)
Volume			
Cubic meters	×	35.314	= cubic feet
Cubic meters	×	1.308	= cubic yards
Cubic meters	×	264.2	= gallons (231 cubic inches)
Liters	×	61.023	= cubic inches (Act. Cong.)
Liters	×	33.82	= fluid ounces (U.S.J.)
Liters	×	0.2642	= gallons (231 cubic inches)
Liters	÷	3.785	= gallons (231 cubic inches)
Liters	÷	28.317	= cubic feet
Hectoliters	×	3.531	= cubic feet
Hectoliters	×	2.838	= bushels (2150.42 cubic inches)
Hectoliters	×	0.1308	= cubic yards
Hectoliters	×	26.42	= gallons (231 cubic inches)
Weight			
Grams	×	15.432	= grains (Act. Cong.)
Grams	÷	981	= dynes
Grams (water)	÷	29.57	= fluid ounces
Grams	÷	28.35	= ounces avoirdupois
Kilograms	×	2.2046	= pounds
Kilograms	×	35.27	= ounces avoirdupois

(continued)

Metric Conversion Table *(continued)*

Weight			
Kilograms	×	0.0011023	= tons (2000 pounds)
Tonneau (Metric ton)	×	1.1023	= tons (2000 pounds)
Tonneau (Metric ton)	×	2204.6	= pounds
Unit Weight			
Grams per cubic centimeter	÷	27.68	= pounds per cubic inch
Kilogram per meter	×	0.672	= pounds per foot
Kilogram per cubic meter	×	0.06243	= pounds per cubic foot
Kilogram per cheval	×	2.235	= pounds per horsepower
Grams per liter	×	0.06243	= pounds per cubic foot
Pressure			
Kilograms per square centimeter	×	14.223	= pounds per square inch
Kilograms per square centimeter	×	32.843	= feet of water (60°F)
Atmospheres (International)	×	14.696	= pounds per square inch
Energy			
Joule	×	0.7376	= foot-pounds
Kilogram-meters	×	7.233	= foot-pounds
Power			
Cheval vapeur	×	0.9863	= horsepower
Kilowatts	×	1.341	= horsepower
Watts	÷	746.	= horsepower
Watts	×	0.7373	= foot-pounds per second

Standard Tables of Metric Measure—Linear Measure

Unit	*Value, m*	*Symbol or Abbreviation*
Micron	0.000001	μ
Millimeter	0.001	mm
Centimeter	0.01	cm
Decimeter	0.1	dm
Meter (unit)	1.0	m
Decameter	10.0	dcm
Hectometer	100.0	hm
Kilometer	1,000.0	km
Myriameter	10,000.0	Mm
Megameter	1,000,000.0	

Volume

Unit	*Value, l*	*Symbol or Abbreviation*
Milliliter	0.001	ml
Centiliter	0.01	cl
Deciliter	0.1	dl
Liter (unit)	1.0	l
Decaliter	10.0	dcl
Hectoliter	100.0	hl
Kiloliter	1,000.0	kl

Surface Measure

Unit	*Value, m²*	*Symbol or Abbreviation*
Square millimeter	0.000001	mm^2
Square centimeter	0.0001	cm^2
Square decimeter	0.01	dm^2
Square meter (centiarc)	1.0	m^2
Square decameter (are)	100.0	a^2
Hectare	10,000.0	ha^2
Square kilometer	1,000,000.0	km^2

(continued)

Standard Tables of Metric Measure—Linear Measure *(continued)*

Mass

Unit	*Value, g*	*Symbol or Abbreviation*
Microgram	0.000001	μg
Milligram	0.001	mg
Centigram	0.01	cg
Decigram	0.1	dg
Gram (unit)	1.0	g
Decagram	10.0	dcg
Hectogram	100.0	hg
Kilogram	1,000.0	kg
Myriagram	10,000.0	Mg
Quintal	100,000.0	q
Ton	1,000,000.0	

Cubic Measure

Unit	*Value, m^3*	*Symbol or Abbreviation*
Cubic micron	10^{-10}	μ^3
Cubic millimeter	10^{-9}	mm^3
Cubic centimeter	10^{-6}	cm^3
Cubic decimeter	10^{-3}	dm^3
Cubic meter	1	m^3
Cubic decameter	10^3	dcm^3
Cubic hectometer	10^6	hm^3
Cubic kilometer	10^9	km^3

Decimal and Millimeter Equivalents of Fractional Parts of an Inch

Parts of Inch	*Decimal*	*Millimeters*	*Parts of Inch*	*Decimal*	*Millimeters*
1/64	0.01563	0.397	33/64	0.51563	13.097
1/32	0.03125	0.794	17/32	0.53125	13.097
3/64	0.04688	1.191	35/64	0.54688	13.890
1/16	0.0625	1.587	9/16	0.5625	14.287
5/64	0.07813	1.984	37/64	0.57813	14.684
3/32	0.09375	2.381	19/32	0.59375	15.081
7/64	0.10938	2.778	39/64	0.60938	15.478
1/8	0.125	3.175	5/8	0.625	15.875
9/64	0.14063	3.572	41/64	0.64063	16.272
3/32	0.15625	3.969	21/32	0.65625	16.669
11/64	0.17188	4.366	43/64	0.67188	17.065
3/16	0.1875	4.762	11/16	0.6875	17.462
13/64	0.20313	5.159	45/64	0.70313	17.859
7/32	0.21875	5.556	23/32	0.71875	18.256
15/64	0.23438	5.953	47/64	0.73438	18.653
1/4	0.25	6.350	3/4	0.75	19.050
17/64	0.26563	6.747	49/64	0.76563	19.447
9/32	0.28125	7.144	25/32	0.78125	19.844
19/64	0.29688	7.541	51/64	0.79688	20.240
5/16	0.3125	7.937	13/16	0.8125	20.637
21/64	0.32813	8.334	53/64	0.82813	21.034
11/32	0.34375	8.731	27/32	0.84375	21.431
23/64	0.35938	9.128	55/64	0.85938	21.828
3/8	0.375	9.525	7/8	0.875	22.225
25/64	0.39063	9.922	57/64	0.89063	22.622
13/32	0.40625	10.319	29/32	0.90625	23.019
27/64	0.42188	10.716	59/64	0.92188	23.415
7/16	0.4375	11.113	15/16	0.9375	23.812
29/64	0.45313	11.509	61/64	0.95313	24.209
15/32	0.46875	11.906	31/32	0.96875	24.606
31/64	0.48438	12.303	63/64	0.98438	25.003
1/2	0.5	12.700	1	1.00000	25.400

BASIC ELECTRICITY

Electrical Terminology

Alternating current (AC). A flow of electricity that constantly reverses or alternates its direction, resulting in a regularly pulsating current.

Amperage. The quantity and rate of flow in an electrical system. It may be compared to the volume of flow in a hydraulic system.

Ampere (A). The unit of electrical current. One ampere is the current that one volt can send through a resistance of one ohm.

Current. The movement or flow of electricity. Current is stated in amperes.

Cycle. The interval or period during which alternating current (using zero as a starting point) increases to maximum force in a positive direction, reverses and decreases to zero, then increases to maximum force in a negative direction, then reverses again and decreases to zero value.

Direct current (DC). A flow of electricity in one direction.

Electromotive force (emf). The force that causes electricity to flow when there is a difference of potential between two points.

Frequency. The number of complete cycles per second of the alternating flow. The most widely used alternating current frequency is 60 cycles per second. This is the number of complete cycles per second. The unit of frequency is the hertz. The term *hertz* (Hz) is defined as cycles per second.

Kilovolt ampere (kVA). One kilovolt ampere is equal to 1000 volt amperes.

Kilowatt (kW). One kilowatt is equal to 1000 watts.

Kilowatt hour (kWH). One kilowatt hour is the electrical energy expended at the rate of one kilowatt (1000 watts) over a period of one hour.

Ohm (Ω). The unit of resistance. One ohm is the resistance offered to the passage of one ampere when moved by one volt.

Phase. The number of current surges that flow simultaneously in an electrical current. In the graphic representation of a single-phase alternating current, the single line represents a current flow that is continuously increasing or decreasing in value. A three-phase circuit has three separate surges of current flowing together. In any given moment, however, their values differ, as the peaks and valleys of the pulsations are spaced equally apart. The wave forms are lettered A, B, and C to represent the alternating current flow for each phase during a complete cycle. In three-phase current flow, any one current pulse is always one-third of a cycle out of matching with another.

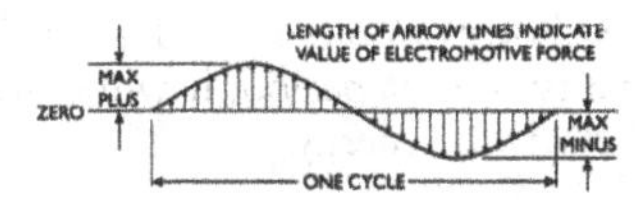

Single-phase alternating current.

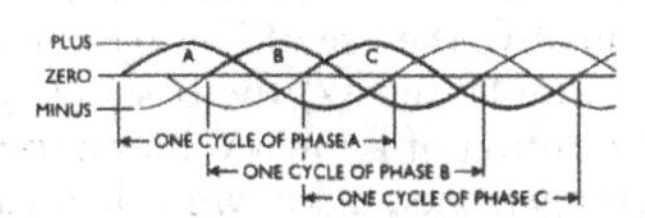

Three-phase alternating current.

Resistance. The resistance offered by materials to the movement of electrons (commonly referred to as the flow of electricity).

Volt (V). The unit of electrical potential or motive force. The force required to send one ampere of current through one ohm of resistance.

Voltage. The value of the electromotive force in an electrical system. It may be compared to pressure in a hydraulic system.

Volt amperes (VA). The product of the volts and amperes measured by a voltmeter and ammeter. In direct current (DC) systems, volt amperes is the same as watts or the energy delivered. In alternating current (AC) systems, the volts and amperes may or may not be 100 percent synchronous. When synchronous, the volt amperes equal the watts. When not synchronous, volt amperes exceed watts.

Watt (W). The electrical unit of power, or the rate of doing work. One watt represents the power used when one ampere of current flows on an electrical circuit with a voltage or pressure of one volt.

Watt hour. The watt hour expresses watts in time measurement hours. For example, if a 100-watt lamp is in operation for a two-hour period, it will consume 200 watt hours of electrical energy.

Electrical Formulas

Most simple electrical calculations associated with common electrical power circuits involve the use of two basic formulas: (1) the *Ohm's law* formula and (2) the basic *electrical power* formulas. By substitution of known values into these formulas, and their rearrangements, unknown values can be easily determined.

Ohm's Law

This is the universally used electrical law stating the relationship of current, voltage, and resistance. This is done mathematically by the formula shown here. Current is stated in *amperes* and abbreviated I. Resistance is stated in *ohms* and abbreviated R, and voltage in *volts* and abbreviated E.

$$\text{Current} = \frac{\text{Voltage}}{\text{Resistance}} \text{ or } I = \frac{E}{R}$$

The arrangement of values gives two other forms of the same equation:

$$R = \frac{E}{I} \text{ and } E = I \times R$$

Example

An ammeter placed in a 110-volt circuit indicates a current flow of 5 amps; what is the resistance of the circuit?

$$R = \frac{E}{I} \text{ or } R = \frac{110}{5} \text{ or } R = 22 \text{ ohms}$$

Power Formulas

This formula indicates the rate at any given instant at which work is being done by current moving through a circuit. Voltage and amperes are abbreviated E and I as in Ohm's law, and watts are abbreviated W.

$$\text{Watts} = \text{volts} \times \text{amperes or } W = E \times I$$

The two other forms of the formula, obtained by rearrangement of the values, are these:

$$E = \frac{W}{I} \text{ or } I = \frac{W}{E}$$

Example

Using the same values as used in the preceding example, 5 amps flowing in a 110-volt circuit, how much power is consumed?

$$W = E \times I \text{ or } W = 110 \times 5 \text{ or } W = 550 \text{ watts}$$

Example

A 110-volt appliance is rated at 2000 watts; can this appliance be plugged into a circuit fused at 15 amps?

$$I = \frac{W}{E} \text{ or } I = \frac{2000}{110} \text{ or } I = 18.18 \text{ amps}$$

Obviously, the fuse would blow if this appliance were plugged into the circuit.

Electric resistance can be calculated by using the following formulas.

$R = E/I$

Resistance (ohms) = voltage (volts) divided by current (amperes)

Or

$R = E^2/W$

Resistance (ohms) = voltage (volts) squared divided by power (watts)

Circuit Basics

An electrical circuit is composed of conductors or conducting devices such as lamps, switches, motors, resistors, wires, cables, batteries, or other voltage sources. Lines and symbols are used to represent the elements of a circuit on paper. These are called *schematic diagrams*. The symbols used to represent the circuit elements, including the voltage source, are standardized. The following table shows the symbols commonly used in industrial applications.

Electrical Symbols

Symbol	Meaning	Symbol	Meaning
	Crossing of conductors not connected		Knife switch
	Crossing of conductors connected		Double-throw switch
	Joining of conductors not crossing		Cable termination
	Grounding connection		Resistor
	Plug connection		Reactor or coil
	Contact normally open		Transformer
	Contact normally closed		Battery
	Fuse	A	Ammeter
	Air circuit breaker	V	Voltmeter
	Oil circuit breaker		

Electrical circuits may be classified as *series* circuits, *parallel* circuits, or a combination of series and parallel circuits. In series circuit, all parts of the circuit are electrically connected end to end. The current flows from one terminal of the power source through each element and to the other power-source terminal. The same amount of current flows in each part of the circuit.

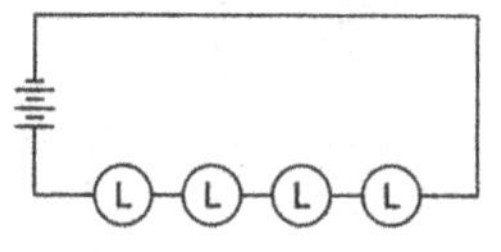

Series circuit.

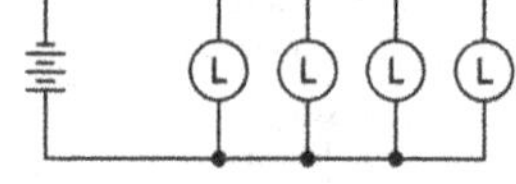

Parallel circuit.

In a parallel circuit, each element is so connected that it has direct flow to both terminals of the power source. The voltage across any element in a parallel circuit is equal to the voltage of the source, or power supply.

The relationship of values in series and parallel circuits using Ohm's law and the power formula are illustrated in and compared in the following examples:

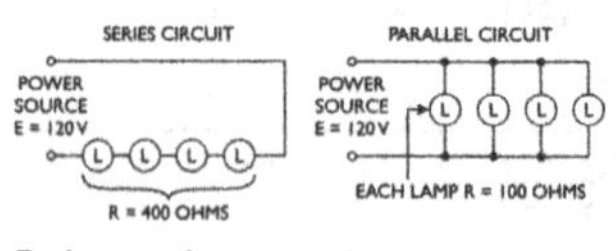

Relationship in values in series and parallel circuits.

Current flow through series circuit	$I = \frac{E}{R} = \frac{120}{400} = 0.3$ amp
Current flow through parallel circuit	$I = \frac{E}{R} = \frac{120}{25} = 4.8$ amps
Voltage across one lamp	$E = IR = 0.3 \times 100 = 30$ volts
Current flow through one lamp	$I = \frac{E}{R} = \frac{120}{100} = 1.2$ amps
Current flow through one lamp	$I = \frac{E}{R} = \frac{30}{100} = 0.3$ amp
Voltage across one lamp	$E = IR = 1.2 \times 100 = 120$ volts
Power used by one lamp	$W = EI = 30 \times 0.3 = 9$ watts
Power used by one lamp	$W = EI = 120 \times 1.2 = 144$ watts
Power used by circuit	$W = EI = 120 \times 0.3 = 36$ watts
Power used by circuit	$W = EI = 120 \times 4.8 = 576$ watts

Electrical Wiring

The term *electrical wiring* is applied to the installation and assembly of electrical conductors. The size of the wire used for electrical conductors is specified by gauge number according to the American Wire Gauge (AWG) system. The usual manner of designation is by the abbreviation AWG. The AWG table that follows lists the AWG numbers and corresponding specifications using the *mil* unit to designate a 0.001-inch measurement.

AWG Table

Size of Wire, AWG	*Diameter of Wire, mils*	*Cross Section, Circular, mils*	*Resistance, ohms/1000 ft at 68°F (20°C)*	*Weight, pounds per 1000 ft*
0000	460	212,000	0.0500	641
000	410	168,000	0.062	508
00	365	133,000	0.078	403
0	325	106,000	0.098	319
1	289	83,700	0.124	253
2	258	66,400	0.156	201
3	229	52,600	0.197	159
4	204	41,700	0.248	126
5	182	33,100	0.313	100
6	162	26,300	0.395	79.5
7	144	20,800	0.498	63.0
8	128	16,500	0.628	50.0
9	144	13,100	0.792	39.6
10	102	10,400	0.998	31.4
11	91	8,230	1.26	24.9
12	81	6,530	1.59	19.8
13	72	5,180	2.00	15.7
14	64	4,110	2.53	12.4
15	57	3,260	3.18	9.86
16	51	2,580	4.02	7.82
17	45	2,050	5.06	6.20
18	40	1,620	6.39	4.92
19	36	1,290	8.05	3.90
20	32	1,020	10.15	3.09
21	28.5	810	12.80	2.45
22	25.3	642	16.14	1.94
23	22.6	509	20.36	1.54
24	20.1	404	25.67	1.22
25	17.9	320	32.37	0.970
26	15.9	254	40.81	0.769
27	14.2	202	51.47	0.610
28	12.6	160	64.90	0.484
29	11.3	127	81.83	0.384
30	10.0	101	103.2	0.304
31	8.9	79.7	130.1	0.241
32	8.0	63.2	164.1	0.191

Current Capacities

Wire Size	In Conduit or Cable: Type RHW*	In Conduit or Cable: Type TW, R*	In Free Air: Type RHW*	In Free Air: Type TW, R*	Weatherproof Wire
14	15	15	20	20	30
12	20	20	25	25	40
10	30	30	40	40	55
8	45	40	65	55	70
6	65	55	95	80	100
4	85	70	125	105	130
3	100	80	145	120	150
2	115	95	170	140	175
1	130	110	195	165	205
0	150	125	230	195	235
00	175	145	265	225	275
000	200	165	310	260	320

** Types RHW, TW, and R are identified by markings on outer cover.*

Adequate Wire Sizes

Load in Building, A	Distance, in ft, from Pole to Building	Recommended* Size of Feeder Wire for Job
Up to 25 A, 120 V	Up to 50	No. 10
	50 to 80	No. 8
	80 to 125	No. 6
20 to 30 A, 240 V	Up to 80	No. 10
	80 to 125	No. 8
	125 to 200	No. 6
	200 to 350	No. 4
30 to 50 A, 240 V	Up to 80	No. 8
	80 to 125	No. 6
	125 to 200	No. 4
	200 to 300	No. 2
	300 to 400	No. 1

** These sizes are recommended to reduce "voltage drop" to a minimum.*

Circuit Wire Sizes for Individual Single-Phase Motors

Horsepower of Motor	*Volts*	*Approximate Starting Current, A*	*Approximate Full Load Current, A*	*Length of Run, in ft. from Main Switch to Motor,*								
				Feet	*25*	*50*	*75*	*160*	*150*	*200*	*300*	*400*
1/4	120	20	5	Wire Size	14	14	14	12	10	10	8	6
1/3	120	20	5.5	Wire Size	14	14	14	12	10	8	6	6
1/2	120	22	7	Wire Size	14	14	12	12	10	8	6	6
3/4	120	28	9.5	Wire Size	14	12	12	10	8	6	4	4
1/4	240	10	2.5	Wire Size	14	14	14	14	14	14	12	12
1/3	240	10	3	Wire Size	14	14	14	14	14	14	12	10
1/2	240	11	3.5	Wire Size	14	14	14	14	14	12	12	10
3/4	240	14	4.7	Wire Size	14	14	14	14	14	12	10	10
1	240	16	5.5	Wire Size	14	14	14	14	14	12	10	10
1 1/2	240	22	7.6	Wire Size	14	14	14	14	12	10	8	8
2	240	30	10	Wire Size	14	14	14	12	10	10	8	6
3	240	42	14	Wire Size	14	12	12	12	10	8	6	6
5	240	69	23	Wire Size	10	10	10	8	8	6	4	4
7 1/2	240	100	34	Wire Size	8	8	8	8	6	4	2	2
10	240	130	43	Wire Size	6	6	6	6	4	4	2	1

Voltage Drop for Copper Conductor

Copper Conductor—90% Factor					
AWG	**Single Phase**	**Three Phase**	**AWG**	**Single Phase**	**Three Phase**
14	0.4762	0.4167	1/0	0.0269	0.0231
12	0.3125	0.2632	2/0	0.0222	0.0196
10	0.1961	0.1677	3/0	0.0190	0.0163
8	0.1250	0.1087	4/0	0.0161	0.0139
6	0.0833	0.0714	250	0.0147	0.0128
4	0.0538	0.0463	300	0.0131	0.0114
3	0.0431	0.0379	350	0.0121	0.0106
2	0.0323	0.0278	400	0.0115	0.0091
1	0.0323	0.0278	500	0.0101	0.0088

Voltage drop formula:

$$\text{Voltage drop} = \frac{90\%\ \text{power factor} \times \text{amps} \times \text{feet}}{1000}$$

Grounding Data

Rating or Setting of Automatic Overcurrent Device in Circuit Ahead of Equipment, Conduit, etc., Not Exceeding (Amps)	Copper	Aluminum or Copper-Clad Aluminum Size (AWG or kcmil)
15	14	12
20	12	10
30	10	8
40	10	8
60	20	8
100	8	6
200	6	4
300	4	2
400	3	1
500	2	1/0
600	1	2/0
800	1/0	3/0
1000	2/0	4/0
1200	3/0	250
1600	4/0	350
2000	250	400
2500	350	600
3000	400	600
4000	500	800
5000	700	1200
6000	800	1200

Use an equipment grounding conductor sized larger than indicated by the preceding table when necessary to comply with Section 250.2(D) of the National Electrical Code. Aluminum or copper-clad aluminum wire comply with the installation restrictions in Section 250.120 of the National Electrical Code.

HEATING FUELS

Fuel Cost Comparison Formulas

To compare various fuel costs on an equal basis, use the following formulas. The costs are based on 100,000 Btu for each fuel.

(1 therm = 100,000 Btu)

Natural Gas

$$\frac{\text{price per cubic ft} \times 100{,}000 \text{ Btu}}{1028 \text{ Btu per cubic foot}} = \text{cost per } 100{,}000 \text{ Btu}$$

Example

$$\frac{0.00629 \times 100{,}000}{1028} = 61.18 \text{ cents per } 100{,}000 \text{ Btu}$$

Note that gas is normally purchased in therms (100,000 Btu) or MCF (1000 cubic feet).

Propane LP Gas

$$\frac{\text{price per gallon} \times 100{,}000 \text{ Btu}}{91{,}333 \text{ Btu per gallon}} = \text{cost per } 100{,}000 \text{ Btu}$$

Example

$$\frac{0.98 \times 100{,}000}{91{,}333} = \$1.07 \text{ per } 100{,}000 \text{ Btu}$$

#2 Fuel Oil

$$\frac{\text{price per gallon} \times 100{,}000 \text{ Btu}}{138{,}690 \text{ Btu per gallon}} = \text{cost per } 100{,}000 \text{ Btu}$$

Example

$$\frac{0.99 \times 100{,}000}{138{,}690} = \$0.71 \text{ per } 100{,}000 \text{ Btu}$$

Electric (kWH)

$$\frac{\text{price per kWH} \times 100{,}000 \text{ Btu}}{3412 \text{ Btu per kWH}} = \text{cost per } 100{,}000 \text{ Btu}$$

Example

$$\frac{0.0831 \times 100{,}000}{3412} = \$2.43 \text{ per } 100{,}000 \text{ Btu}$$

The "apples to apples" fuel cost comparisons generated by these calculations are the starting point for further cost analysis. To get a true picture of actual fuel costs, one must consider the specific furnace or heat-pump operating efficiencies plus the heating demands of the particular building and the geographical location of that building.

Average Btu Content of Common Heating Fuels

Fuel Type	*Number of Btu/Unit*
Fuel oil (No. 2)	140,000/gallon
Natural gas	1,025,000/thousand cubic feet
Propane	91,330/gallon
Coal	28,000,000/ton
Electricity	3,412/kWh
Wood (air dried)	20,000,000/cord or 8000/pound
Kerosene	135,000/gallon
Pellets	16,500,000/ton

Courtesy Office of Energy Efficiency and Renewable Energy/U.S. Department of Energy

Gas Consumption of Typical Appliances

Appliance	*Input Btu/hr (approx.)*
Boiler or furnace (domestic)	100,000 to 250,000
Range (freestanding, domestic)	65,000
Built-in oven or broiler unit (domestic)	25,000
Built-in top unit (domestic)	40,000
Water heater, automatic storage (50-gal tank)	55,000
Water heater, automatic instantaneous	
2 gal per minute	142,800
4 gal per minute	285,000
6 gal per minute	428,400
Water heater, circulating or side-arm (domestic)	35,000
Refrigerator	3,000
Clothes dryer (domestic)	35,000

Estimated Average Fuel Conversion Efficiency

Fuel Type	*Appliance Type*	*Fuel Efficiency*
Bituminous coal	Hand-fired central heating appliance	45.0
	Stoker-fired central heating appliance	60.0
Oil	High-efficiency central heating appliance	89.0
	Typical central heating appliance	80.0
	Water heater (50 gal)	59.5
Gas	High-efficiency central heating furnace	97.0
	Typical central heating boiler	85.0
	Minimum efficiency central heating furnace	78.0
	Room heater (unvented)	99.0
	Room heater (vented)	65.0
	Water heater (50 gal)	62.0
Electricity	Baseboard (resistance)	99.0
	Forced-air central heating appliance	97.0
	Heat pump central heating system	200+
	Ground source heat pump	300+
	Water heater (50 gal)	97.0
Wood and pellets	Franklin stove	30.0–40.0
	Stoves with circulating fans	40.0–70.0
	Catalytic stoves	65.0–75.0
	Pellet stoves	85.0–90.0

Courtesy Office of Energy Efficiency and Renewable Energy/U.S. Department of Energy.

Propane and Butane Btu/Weight Comparisons

Property	*Butane*	*Propane*
Btu per ft³ 60°F	3280	2516
Btu per lb	21,221	21,591
Btu per gal	102,032	91,547
Ft³ per lb	6.506	8.58
Ft³ per gal	31.26	36.69
Lb per gal	4.81	4.24

Courtesy National LP-Gas Association

Physical and Chemical Properties of Propane

Boiling point at 14.7 psia =	–44°F
Specific gravity of vapor (air = 1) at 60°F =	1.50
Specific gravity of liquid (water = 1) at 60°F =	0.504
Vapor pressure at 70°F =	127 psig
Vapor pressure at 105°F =	210 psig
Expansion ratio (from liquid to gas) at 14.7 psia =	1 to 270
Solubility in water	Slight, 0.1 to 1.0%
Appearance and odor	Colorless and tasteless gas at normal temperature and pressure*

To aid in the detection of leaks, an odorant (commonly ethyl mercaptan) is added to propane to provide a strong unpleasant odor. Unfortunately, no odorant additive is 100 percent effective. The odor level may be reduced by such factors as chemical reactions with material in the system, or leaking propane passing through some kinds of soils. If the odor of the propane seems to be weaker than normal, immediately contact the local propane supplier and ask to have the system checked.

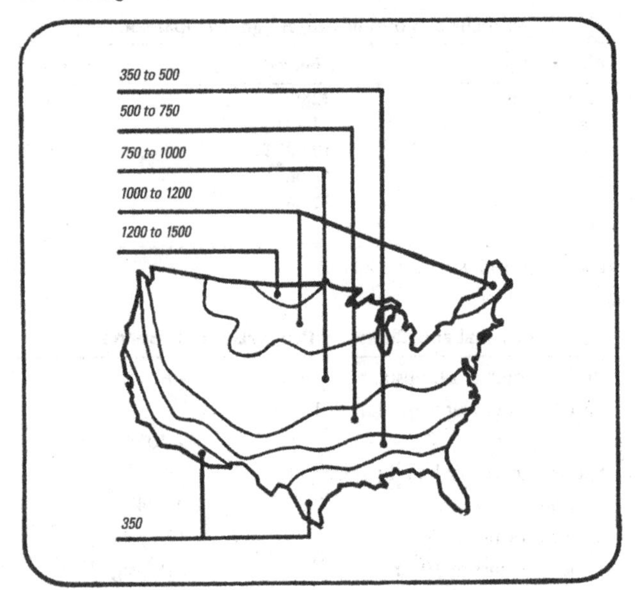

The six basic weather zones and the propane tank size required to ensure an adequate supply of gas in each. *(Courtesy National Propane Gas Association)*

The construction of residential propane storage tanks is governed by state and local regulations. The material of the tank will meet minimum ASME design criteria, the local code requirement, and the proposed service. Most are designed with a maximum design temperature of 125°F and a maximum design pressure of 250 psig.

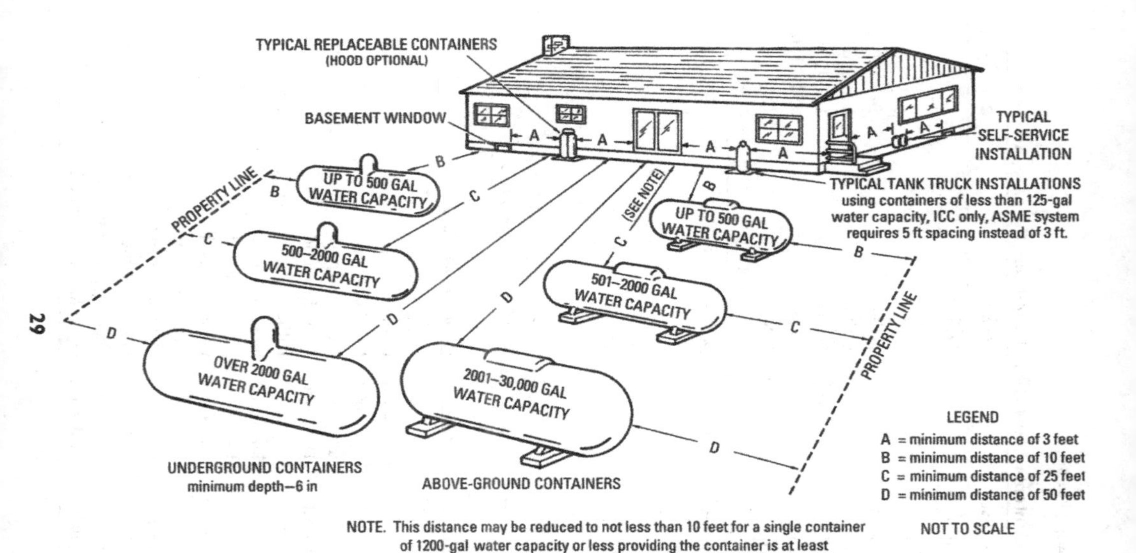

Suggested clearances for LP gas cylinders and storage tanks. *(Courtesy National Propane Gas Association)*

Btu Content of Fuel Oils, Natural Gas, Propane, and Butane

Grade or Type	*Unit*	*Btu*
No. 1 oil	Gallon	137,400
No. 2 oil	Gallon	139,600
No. 3 oil	Gallon	141,800
No. 4 oil	Gallon	145,100
No. 5 oil	Gallon	148,800
No. 6 oil	Gallon	152,400
Natural gas	Cubic foot	950–1150
Propane	Cubic foot	2550
Butane	Cubic foot	3200

Courtesy Honeywell Tradeline Controls

Combustion Characteristics of Typical Firewoods

Name	*Type*	*Combustion Characteristics*
Ash, white	Hardwood	Good firewood
Beech	Hardwood	Good firewood
Birch, yellow	Hardwood	Good firewood
Chestnut	Hardwood	Excessive sparking (can be dangerous)
Cottonwood	Hardwood	Good firewood
Elm, white	Hardwood	Difficult to split, but burns well
Hickory	Hardwood	Slow, steady fire; best firewood
Maple, sugar	Hardwood	Good firewood
Maple, red	Hardwood	Good firewood
Oak, red	Hardwood	Slow, steady fire
Oak, white	Hardwood	Slow, steady fire
Pine, yellow	Softwood	Quick, hot fire; smokier than hardwood
Pine, white	Softwood	Quick, hot fire; smokier than hardwood
Walnut, black	Hardwood	Good firewood, but difficult to find

Heat Value (Million Btu per Cord) of Wood with 12% Moisture Content

Name of Wood	*Heat Value*	*Equivalent Coal Heat Value*
Ash, white	28.3	1.09
Beech	31.1	1.20
Birch, yellow	30.4	1.17
Chestnut	20.7	0.80
Cottonwood	19.4	0.75
Elm, white	24.2	0.93
Hickory	35.3	1.36
Maple, sugar	30.4	1.17
Maple, red	26.3	1.01
Oak, red	30.4	1.17
Oak, white	32.5	1.25
Pine, yellow	26.0	1.00
Pine, white	18.1	0.70

**The equivalent coal heat values are based on 1 ton (2000 lb) of anthracite coal with a heat value of 13,000 Btu/lb.*

Ignition Temperatures of Coal and Wood

Type	*Ignition Temperature*
Paper	350°F
Wood	435°F
Western lignite coal	630°F
Low volatile bituminous coal	765°F
High volatile bituminous coal	870°F
Anthracite coal	925°F

Courtesy Yukon/Eagle

Btu Per Cord of Air-Dried Wood

Type	*Pound Weight Per Cord*	*Btu Per Cord*	*Equivalent Value #2 Fuel Oil (Gal)*
White Pine	1800	17,000,000	120
Aspen	1900	17,500,000	125
Spruce	2100	18,000,000	130
Ash	2900	22,500,000	160
Tamarack	2500	24,000,000	170
Soft Maple	2500	24,000,000	170
Yellow Birch	3000	26,000,000	185
Red Oak	3250	27,000,000	195
Hard Maple	3000	29,000,000	200
Hickory	3600	30,500,000	215

Courtesy Yukon/Eagle

PIPING AND PIPEFITTING

Symbols for Pipe Fittings Commonly Used in Drafting

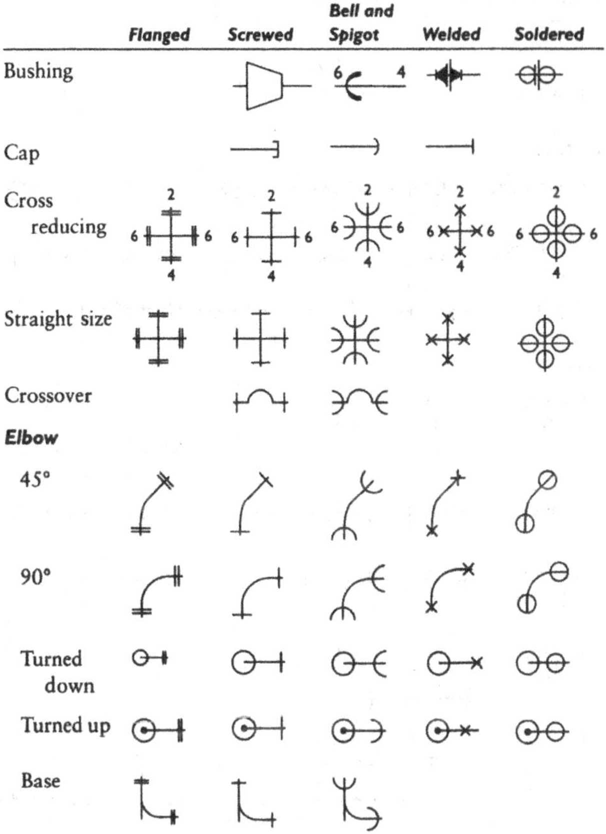

	Flanged	*Screwed*	*Bell and Spigot*	*Welded*	*Soldered*
Bushing			6 4		
Cap					
Cross reducing	2 6 6 4	2 6 6 4	2 6 6 4	2 6 6 4	2 6 6 4
Straight size					
Crossover					
Elbow					
45°					
90°					
Turned down					
Turned up					
Base					

	Flanged	Screwed	Bell and Spigot	Welded	Soldered
Plugs					
Bull plug					
Pipe plug					
Reducer					
Concentric					
Eccentric					
Sleeve					
Tee					
Straight size					
(Outlet Up)					
(Outlet Down)					
Double sweep					
Reducing	2 6 4	2 6 4	2 6 4	2 6 4	2 6 4

American Standard Pipe Thread Standards

Standard Pipe Threads

	A	B	E	F	G	H		P	
Nominal Size of Pipe, in.	Pitch Dia. at End of Pipe, in.	Pitch Dia. at Gauging Notch, in.	Length of Effective Thread, in.	Normal Engagement by Hand Between Male and Female Thread, in.	Outside Dia. of Pipe, in.	Actual Inside Dia. of Pipe, in.	Number of Threads per Inch,	Pitch of Thread, in.	Depth of Thread, in.
⅛	0.36351	0.37476	0.2638	0.180	0.405	0.269	27	0.0370	0.02963
¼	0.47739	0.48989	0.4018	0.200	0.540	0.364	18	0.0556	0.04444
⅜	0.61201	0.62701	0.4078	0.240	0.675	0.493	18	0.0556	0.04444
½	0.75843	0.77843	0.5337	0.320	0.840	0.622	14	0.0714	0.05714
¾	0.06768	0.98886	0.5457	0.339	1.050	0.824	14	0.0714	0.05714
1	1.21363	1.23863	0.6828	0.400	1.315	1.049	11½	0.0870	0.06954
1¼	1.55713	1.58338	0.7068	0.420	1.660	1.380	11½	0.0870	0.06954
1½	1.79609	1.82234	0.7235	0.420	1.900	1.610	11½	0.0870	0.06956
2	2.26902	2.29627	0.7565	0.436	2.375	2.067	11½	0.0870	0.06956
2½	2.71953	2.76216	1.1375	0.681	2.875	2.469	8	0.1250	0.10000
3	3.34063	3.38850	1.2000	0.766	3.500	3.068	8	0.1250	0.10000
3½	3.83750	3.88881	1.2500	0.821	4.000	3.548	8	0.1250	0.10000
4	4.33438	4.38713	1.3000	0.844	4.500	4.026	8	0.1250	0.10000
4½	4.83125	4.88594	1.3500	0.875	5.000	4.506	8	0.1250	0.10000
5	5.39073	5.44929	1.4063	0.937	5.563	5.047	8	0.1250	0.10000

(continued)

(continued)

	A	B	E	F	G	H		P	
Nominal Size of Pipe, in.	Pitch Dia. at End of Pipe, in.	Pitch Dia. at Gauging Notch, in.	Length of Effective Thread, in.	Normal Engagement by Hand Between Male and Female Thread, in.	Outside Dia. of Pipe, in.	Actual Inside Dia. of Pipe, in.	Number of Threads per Inch,	Pitch of Thread, in.	Depth of Thread, in.
6	6.44609	6.50597	1.5125	0.958	6.625	6.055	8	0.1250	0.10000
7	7.43984	7.50234	1.6125	1.000	7.625	7.023	8	0.1250	0.10000
8	8.43359	8.50003	1.7125	1.063	8.625	7.981	8	0.1250	0.10000
9	9.42734	9.49797	1.8125	1.130	9.625	8.941	8	0.1250	0.10000
10	10.54531	10.62094	1.9250	1.210	10.750	10.020	8	0.1250	0.10000
12	12.53281	12.61781	2.1250	1.360	12.750	12.000	8	0.1250	0.10000
14 OD	13.77500	13.87262	2.250	1.562	14.000	—	8	0.1250	0.10000
15 OD	14.76875	14.87419	2.350	1.687	15.000	—	8	0.1250	0.10000
16 OD	15.76250	15.87575	2.450	1.812	16.000	—	8	0.1250	0.10000
18 OD	17.75000	17.87500	2.650	2.000	18.000	—	8	0.1250	0.10000
20 OD	19.73750	19.87031	2.850	2.125	20.000	—	8	0.1250	0.10000
22 OD	21.72500	21.86562	3.050	2.250	22.000	—	8	0.1250	0.10000
24 OD	23.71250	23.86094	3.250	2.375	24.000	—	8	0.1250	0.10000

Data abstracted from the American Standard for Pipe Threads A.S.A.-B2—1919

American Standard Pipe Thread Designations

American Standard Pipe Threads are designated by specifying in sequence the nominal size, number of threads per inch, and the thread series symbols:

Nominal Size	*No. of Threads*	*Symbols*
3/8	18	NPT

Where:

N—American (Nat.) Standard
P—Pipe
T—Taper
C—Coupling
S—Straight
L—Locknut
R—Railing Fittings
M—Mechanical

Examples:

3/8—18 NPT, American Standard Taper Pipe Thread

3/8—18 NPSC, American Standard Straight Coupling Pipe Thread

1/8—27 NPTR, American Standard Taper Railing Pipe Thread

1/2—14 NPSM, American Standard Straight Mechanical Pipe Thread

1—11 1/2 NPSL, American Standard Straight Locknut Pipe Thread

Left-hand threads are designated by adding LH.

American Standard Taper Pipe Threads (NPT)

Basic Dimensions

Taper pipe threads are engaged or made up in two phases, *band engagement* and *wrench makeup.*

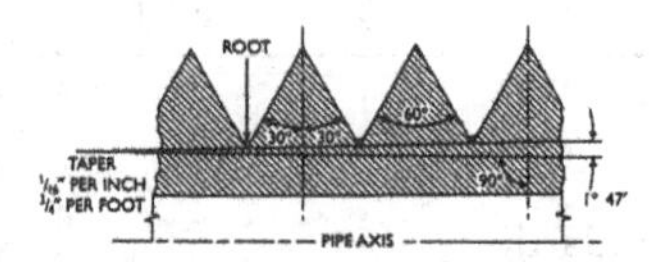

American Standard taper pipe thread form.

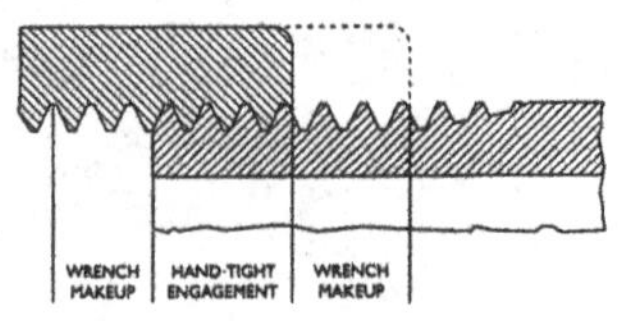

Pipe thread engagement.

Taper Pipe Makeup Dimensions*

		Hand Tight		Wrench Makeup		Dimension	Turns
Pipe Size	Threads per Inch	Dimension	Turns	Dimension	Turns	Total Makeup	
1/8	27	3/16	4 1/2	3/32	2 1/2	9/32	7
1/4	18	7/32	4	3/16	3	13/16	7
3/8	18	1/4	4 1/2	3/16	3	7/16	7 1/2
1/2	14	5/16	4 1/2	7/32	3	17/32	7 1/2
3/4	14	5/16	4 1/2	7/32	3	17/32	7 1/2
1	11 1/2	3/8	4 1/2	1/4	3 1/4	11/16	8
1 1/4	11 1/2	13/32	4 1/2	9/32	3 1/4	11/16	8
1 1/2	11 1/2	7/16	5	1/4	3	11/16	8
2	11 1/2	7/16	5	1/4	3	11/16	8 1/2
2 1/2	8	11/16	5 1/2	3/8	3	1 1/8	9
3	8	3/4	6	3/8	3	1 1/8	9

**This table includes basic makeup dimensions. Commercial products may vary as much as one turn larger or smaller and still be within standard tolerance. In practice, pipe threads are usually cut to give a connection that makes up less than the basic standard. Common practice is about 3 turns by hand and 3 to 4 turns by wrench.*

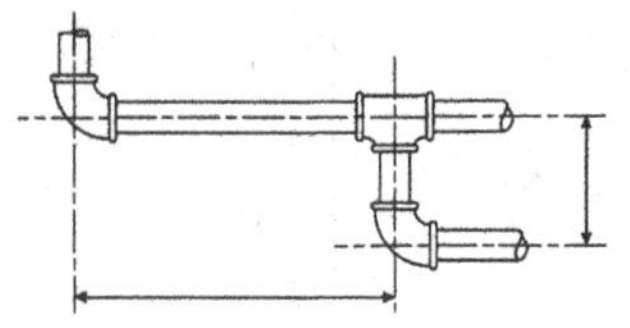

Location of centerlines and/or points on centerlines.

Pipe Measurement and Identification

Dimensions on pipe drawings specify the location of centerlines and/or points on centerlines; they do not specify pipe lengths. This system of distance dimensioning and measurement is also followed in the fabrication and installation of pipe assemblies.

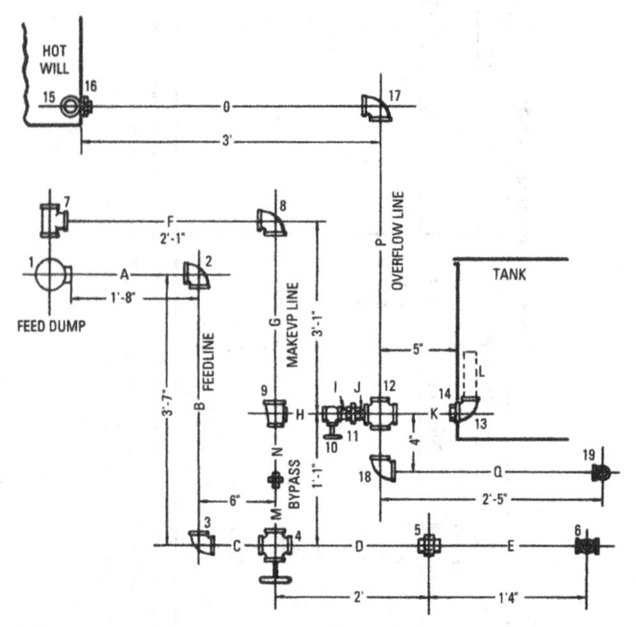

Working drawing showing centerline and distances between centers of an installation.

Takeout Allowances for Screwed Pipe

To determine actual pipe lengths, allowances must be made for the length of the fittings and the distance threaded pipe is made up into the fittings. To do this, subtract an amount called *takeout* from the *center-to-center* dimension. The relationships of takeout to other threaded pipe connection distances, termed *makeup*, *center-to-center*, and *end-to-end*, are illustrated in the figure.

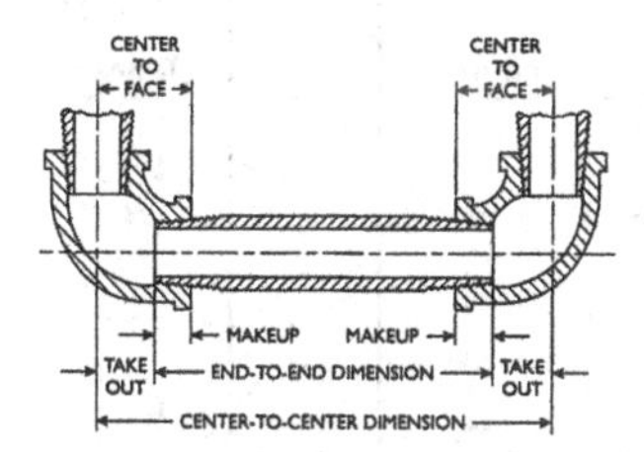

The relationships of takeout to other threaded pipe connection distances.

To determine end-to-end pipe length, the takeout is subtracted from the center-to-center dimension. Standard tables may be used for this purpose. These tables should be used with judgment, however, because commercial product tolerance is one turn plus or minus. On critical connections, materials should be checked and compensation made for variances.

Takeout Allowances

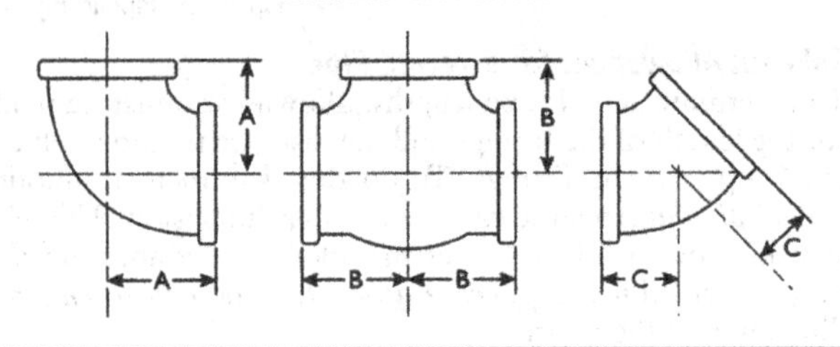

Pipe Size	90° Elbow A	90° Elbow Takeout	Tee B	Tee Takeout	45° Elbow C	45° Elbow Takeout
1/8	11/16	7/16	11/16	7/16	9/16	1/4
1/4	13/16	7/16	13/16	7/16	3/4	3/8
3/8	15/16	9/16	15/16	9/16	13/16	7/16
1/2	1 1/8	5/8	1 1/8	5/8	7/8	3/8
3/4	1 5/16	3/4	1 15/16	3/4	1	7/16
1	1 1/2	7/8	1 1/2	7/8	1 1/3	9/16
1 1/4	1 3/4	1 1/8	1 3/4	1 1/8	1 15/16	11/16
1 1/2	1 15/16	1 1/4	1 15/16	1 1/4	1 7/16	3/4
2	2 1/4	1 5/8	2 1/4	1 5/8	1 11/16	1

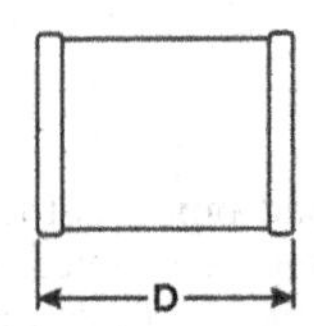

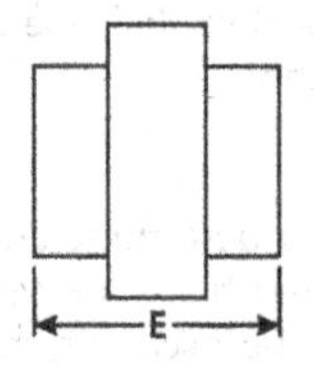

Pipe Size	Thread Makeup	Coupling D	Coupling Takeout	Union E	Union Takeout
1/8	1/4	1	1/4	1 1/2	3/4
1/4	3/8	1 1/8	3/8	1 5/8	7/8
3/8	3/8	1 1/4	3/8	1 5/8	7/8
1/2	1/2	1 3/8	3/8	1 7/8	1
3/4	9/16	1 1/2	3/8	1 7/8	1
1	9/16	1 3/4	1/2	2 3/8	1 1/4
1 1/4	5/8	2	3/4	2 5/8	1 3/8
1 1/2	5/8	2 1/2	7/8	3	1 1/2
2	11/16	2 1/2	1 1/4	3 1/4	1 3/4

Pipe Schedules

Pipe schedule numbers indicate pipe strength. The higher the number in a given size, the greater the strength. The schedule number indicates the approximate values of the following expression:

$$\text{Schedule number} = \frac{1000 \times \text{internal pressure}}{\text{allowable stress in pipe}}$$

Commercial Pipe Sizes and Wall Thickness

Nominal Pipe Size	Outside Diameter	Nominal Wall Thickness: Sched. 5	Sched. 10	Sched. 40 Std.	Sched. 80 Ex. Std.	Sched. 160	Ex. Ex. Strong
1/8	0.405	-	.049	.068	.095	-	-
1/4	0.540	-	.065	.088	.119	-	-
3/8	0.675	-	.065	.091	.126	-	-
1/2	0.840	-	.083	.109	.147	.187	.294
3/4	1.050	.065	.083	.113	.154	.218	.308
1	1.315	.065	.109	.133	.179	.250	.358
1 1/4	1.660	.065	.109	.140	.191	.250	.382
1 1/2	1.900	.065	.109	.145	.200	.281	.400
2	2.375	.065	.109	.154	.218	.343	.436
2 1/2	2.875	.083	.120	.203	.276	.375	.552
3	3.500	.083	.120	.216	.300	.438	.600
3 1/2	4.000	.083	.120	.226	.318	-	-
4	4.500	.083	.120	.237	.337	.531	.674
5	5.563	.109	.134	.258	.375	.625	.750
6	6.625	.109	.134	.280	.432	.718	.864
8	8.625	.109	.148	.322	.500	.906	.875

Schedule 40 Pipe Dimensions

	Diameter		Nominal	Transverse Area			Length of Pipe per ft²				
Size, in.	External, in.	Internal, in.	Thickness, in.	External, in.²	Internal, in.²	Metal, in.²	External Surface, ft	Internal Surface, ft	Cubic Feet per ft of Pipe	Weight per ft	Number Threads per in. of Screw
⅛	0.405	0.269	0.068	0.129	0.057	0.072	9.431	14.199	0.00039	0.244	27
¼	0.54	0.364	0.088	0.229	0.104	0.125	7.073	10.493	0.00072	0.424	18
⅜	0.675	0.493	0.091	0.358	0.191	0.167	5.658	7.747	0.00133	0.567	18
½	0.84	0.622	0.109	0.554	0.304	0.25	4.547	6.141	0.00211	0.85	14
¾	1.05	0.824	0.113	0.866	0.533	0.333	3.637	4.635	0.0037	1.13	14
1	1.315	1.049	0.133	1.358	0.864	0.494	2.904	3.641	0.006	1.678	11½
1¼	1.66	1.38	0.14	2.164	1.495	0.669	2.301	2.767	0.01039	2.272	11½
1½	1.9	1.61	0.145	2.835	2.036	0.799	2.01	2.372	0.01414	2.717	11½
2	2.375	2.067	0.154	4.43	3.355	1.075	1.608	1.847	0.0233	3.652	11½
2½	2.875	2.469	0.203	6.492	4.788	1.704	1.328	1.547	0.03325	5.793	8
3	3.5	3.068	0.216	9.621	7.393	2.228	1.091	1.245	0.05134	7.575	8
3½	4	3.548	0.226	12.56	9.886	2.68	0.954	1.076	0.06866	9.109	8
4	4.5	4.026	0.237	15.9	12.73	3.174	0.848	0.948	0.0884	10.79	8
5	5.563	5.047	0.258	24.3	20	4.3	0.686	0.756	0.1389	14.61	8
6	6.625	6.065	0.28	34.47	28.9	5.581	0.576	0.629	0.2006	18.97	8
8	8.625	7.981	0.322	58.42	50.02	8.399	0.442	0.478	0.3552	28.55	8

(continued)

(continued)

Size, in.	Diameter		Nominal Thickness, in.	Transverse Area			Length of Pipe per ft²		Cubic Feet per ft of Pipe	Weight per ft	Number Threads per in. of Screw
	External, in.	Internal, in.		External, in²	Internal, in²	Metal, in²	External Surface, ft	Internal Surface, ft			
10	10.75	10.02	0.365	90.76	78.85	11.9	0.355	0.381	0.5476	40.48	8
12	12.75	11.938	0.406	127.64	111.9	15.74	0.299	0.318	0.7763	53.6	
14	14	13.125	0.437	153.94	135.3	18.64	0.272	0.28	0.9354	63	
16	16	15	0.5	201.05	176.7	24.35	0.238	0.254	1.223	78	
18	18	16.874	0.563	254.85	224	30.85	0.212	0.226	1.555	105	
20	20	18.814	0.593	314.15	278	36.15	0.191	0.203	1.926	123	
24	24	22.626	0.687	452.4	402.1	50.3	0.159	0.169	2.793	171	

Courtesy Sarco Company, Inc.

Schedule 80 Pipe Dimensions

	Diameter		Nominal	Transverse Area			Length of Pipe per ft²				
Size, in.	External, in.	Internal, in.	Thickness, in.	External, in²	Internal, in²	Metal, in²	External Surface, ft	Internal Surface, ft	Cubic Feet per ft of Pipe	Weight per ft,	Number Threads per in. of Screw
1/8	0.405	0.215	0.095	0.129	0.036	0.093	9.431	17.75	0.00025	0.314	27
1/4	0.54	0.302	0.119	0.229	0.072	0.157	7.073	12.65	0.0005	0.535	18
3/8	0.675	0.423	0.126	0.358	0.141	0.217	5.658	9.03	0.00098	0.738	18
1/2	0.84	0.546	0.147	0.554	0.234	0.32	4.547	7	0.00163	1	14
3/4	1.05	0.742	0.154	0.866	0.433	0.433	3.637	5.15	0.003	1.47	14
1	1.315	0.957	0.179	1.358	0.719	0.639	2.904	3.995	0.005	2.17	11½
1¼	1.66	1.278	0.191	2.164	1.283	0.881	2.301	2.99	0.00891	3	11½
1½	1.9	1.5	0.2	2.835	1.767	1.068	2.01	2.542	0.01227	3.65	11½
2	2.375	1.939	0.218	4.43	2.953	1.477	1.608	1.97	0.02051	5.02	11½
2½	2.875	2.323	0.276	6.492	4.238	2.254	1.328	1.645	0.02943	7.66	8
3	3.5	2.9	0.3	9.621	6.605	3.016	1.091	1.317	0.04587	10.3	8
3½	4	3.364	0.318	12.56	8.888	3.678	0.954	1.135	0.06172	12.5	8
4	4.5	3.826	0.337	15.9	11.497	4.407	0.848	0.995	0.0798	14.9	8
5	5.563	4.813	0.375	24.3	18.194	6.112	0.686	0.792	0.1263	20.8	8
6	6.625	5.761	0.432	34.47	26.067	8.3	0.576	0.673	0.181	28.6	8
8	8.625	7.625	0.5	58.42	46.663	12.76	0.442	0.501	0.3171	43.4	8

(continued)

(continued)

Size, in.	Diameter		Nominal Thickness, in.	Transverse Area			Length of Pipe per ft^2		Cubic Feet per ft of Pipe	Weight per ft,	Number Threads per in. of Screw
	External, in.	Internal, in.		External, in^2	Internal, in^2	Metal, in^2	External Surface, ft	Internal Surface, ft			
10	10.75	9.564	0.593	90.76	71.84	18.92	0.355	0.4	0.4989	64.4	8
12	12.75	11.376	0.687	127.64	101.64	26	0.299	0.336	0.7058	88.6	
14	14	12.5	0.75	153.94	122.72	31.22	0.272	0.306	0.8522	107	
16	16	14.314	0.843	201.05	160.92	40.13	0.238	0.263	1.112	137	
18	18	16.126	0.937	254.85	204.24	50.61	0.212	0.237	1.418	171	
20	20	17.938	1.031	314.15	252.72	61.43	0.191	0.208	1.755	209	
24	24	21.564	1.218	452.4	365.22	87.18	0.159	0.177	2.536	297	

Courtesy Sarco Company, Inc.

Pipe Tapping

An internal (female) thread is cut into a pipe with a pipe tap, a conical screw made of hardened steel with longitudinal grooves. In some cases the pipe may have to be reamed before the female threads are cut. The following table lists drill sizes that permit direct tapping without reaming the holes beforehand. The table lists drill sizes for both the Briggs (American Standard) and the Whitworth (British Standard) pipe taps.

Drill Sizes for Briggs Standard Pipe Taps (For Direct Tapping Without Reaming)

Size of pipe	1/8	1/4	3/8	1/2	3/4	1	1 1/4	1 1/2	2	2 1/2	3	3 1/2	4
Size of drill	21/64	7/16	9/16	45/64	29/32	1 9/64	1 31/64	1 47/64	2 13/64	2 5/8	3 1/4	3 47/64	4 15/64

Drill Sizes for Pipe Taps (Briggs and British Standard)

Size Taps, in.	Briggs Standard		British (Whitworth) Standard	
	Thread	Drill	Thread	Drill
1/8	27	21/64	28	5/16
1/4	18	27/64	19	7/16
3/8	18	9/16	19	9/16
1/2	14	11/16	14	23/32
5/8	—	—	14	25/32
3/4	14	29/32	14	29/32
7/8	—	—	14	1 1/16
1	11 1/2	1 1/8	11	1 5/32
1 1/4	11 1/2	1 15/32	11	1 1/2
1 1/2	11 1/2	1 23/32	11	1 23/32
1 3/4	—	—	11	1 31/32
2	11 1/2	2 3/16	11	2 3/16
2 1/4	—	—	11	2 13/32
2 1/2	8	2 9/16	11	2 25/32
2 3/4	—	—	11	3 1/32
3	8	3 3/16	11	3 9/32
3 1/4	—	—	11	3 1/2
3 1/2	8	3 11/16	11	3 3/4
3 3/4	—	—	11	4
4	8	4 3/16	11	4 1/4
4 1/2	8	4 11/16	11	4 3/4
5	8	5 1/4	11	5 1/4
5 1/2	—	—	11	5 3/4
6	8	6 5/16	11	6 1/4
7	8	7 5/16	11	7 5/16
8	8	8 5/16	11	8 5/16
9	8	9 5/16	11	9 5/16
10	8	10 7/16	11	10 5/16

Pipe Fittings

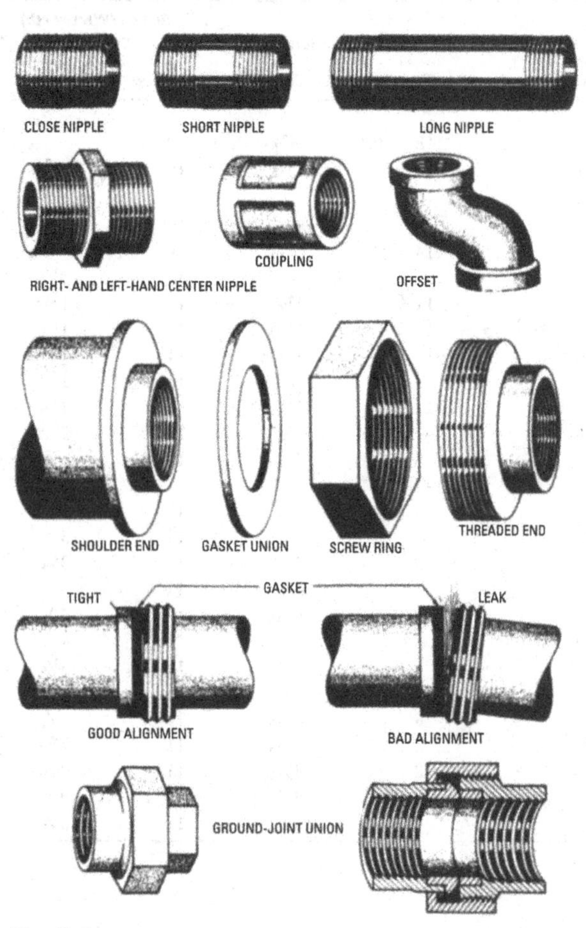

Pipe fittings.

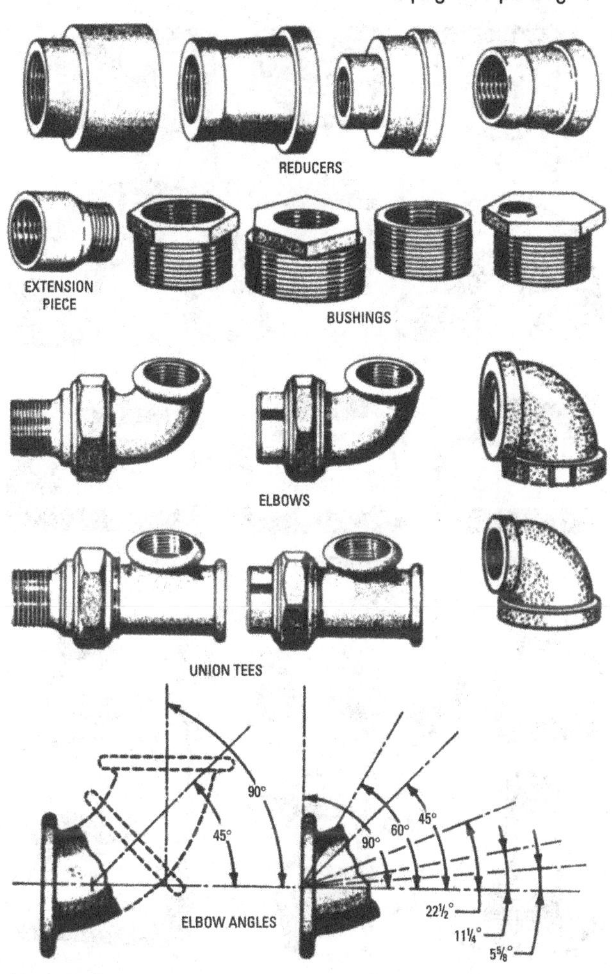
REDUCERS
EXTENSION PIECE
BUSHINGS
ELBOWS
UNION TEES
90°
45°
90°
60°
45°
22½°
11¼°
5⅝°
ELBOW ANGLES

(continued)

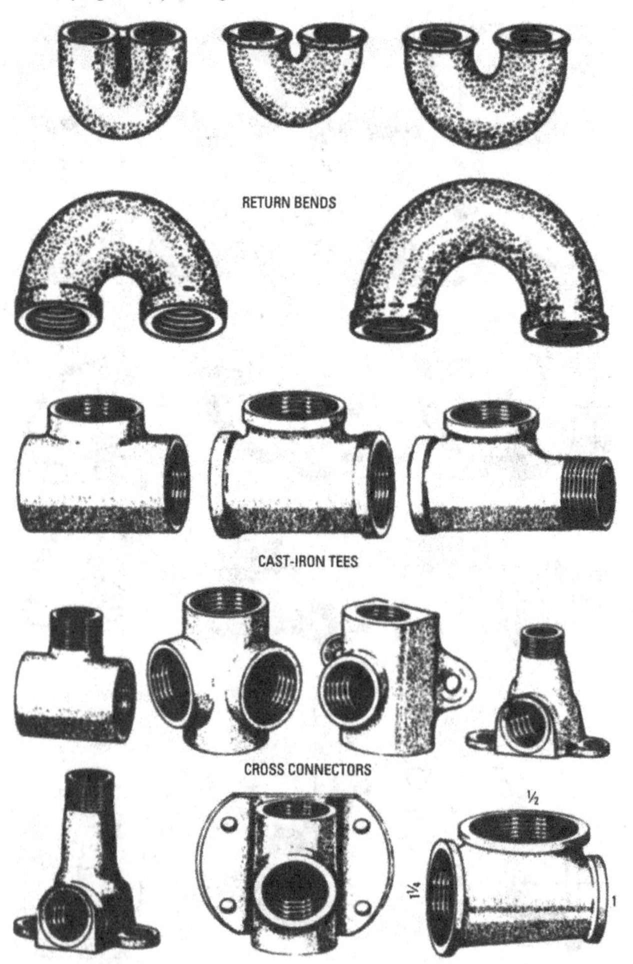

(continued)

Y-BRANCHES

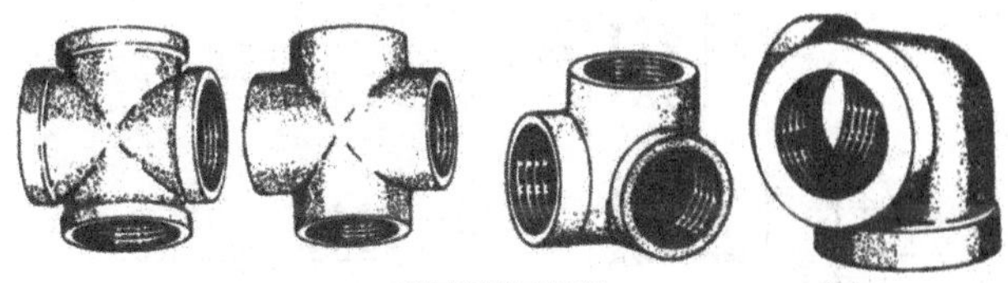

SIDE-OUTLET ELBOWS

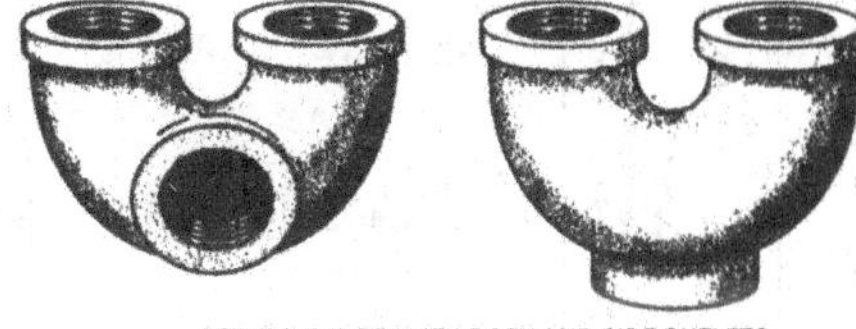

RETURN BENDS WITH BACK AND SIDE OUTLETS

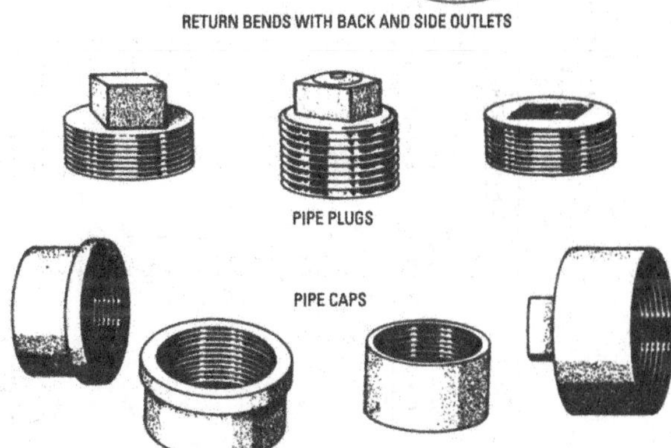

PIPE PLUGS

PIPE CAPS

(continued)

Dimensions of Steel Butt-Welding Fittings

Nominal Pipe Size	Long-Radius Elbows: Outside Diameter at Bevel	Long-Radius Elbows: Center to End: 90° Elbows (A)	Long-Radius Elbows: Center to End: 45° Elbows (B)	180° Returns: Outside Diameter at Bevel	180° Returns: Center to Center (O)	180° Returns: Back to Face (K)	Straight Tees: Outside Diameter at Bevel	Straight Tees: Center to End: Run (C)	Straight Tees: Center to End: Outlet (M)
1	1.315	1 1/2	7/8	1.315	3	2 3/16	1.315	1 1/2	1 1/2
1 1/4	1.660	1 7/8	1	1.660	3 3/4	2 3/4	1.660	1 7/8	1 7/8
1 1/2	1.900	2 1/4	1 1/8	1.900	4 1/2	3 1/4	1.900	2 1/4	2 1/4
2	2.375	3	1 3/8	2.375	6	4 3/16	2.375	2 1/2	2 1/2
2 1/2	2.875	3 3/4	1 3/4	2.875	7 1/2	5 3/16	2.875	3	3
3	3.500	4 1/2	2	3.500	9	6 1/4	3.500	3 3/8	3 3/8
3 1/2	4.000	5 1/4	2 1/4	4.000	10 1/2	7 1/4	4.000	3 3/4	3 3/4
4	4.500	6	2 1/2	4.500	12	8 1/4	4.500	4 1/8	4 1/8
5	5.563	7 1/2	3 1/8	5.563	15	10 5/16	5.563	4 7/8	4 7/8
6	6.625	9	3 3/4	6.625	18	12 5/16	6.625	5 5/8	5 5/8
8	8.625	12	5	8.625	24	16 5/16	8.625	7	7
10	10.750	15	6 1/4	10.750	30	20 3/8	10.750	8 1/2	8 1/2
12	12.750	18	7 1/2	12.750	36	24 3/8	12.750	10	10
14	14.000	21	8 3/4	14.000	42	28	14.000	11	
16	16.000	24	10	16.000	48	32	16.000	12	Not standard
18	18.000	27	11 1/4	18.000	54	36	18.000	13 1/2	
20	20.000	30	12 1/2	20.000	60	40	20.000	15	
24	24.000	36	15	24.000	72	48	24.000	17	

From American Standard for Butt-Welding Fittings, ASA B16.9-1958. All dimensions are in inches. Dimension A is equal to 1/2 of dimension O.

Courtesy 1960 ASHRAE Guide

General Dimensions of Straight Sizes

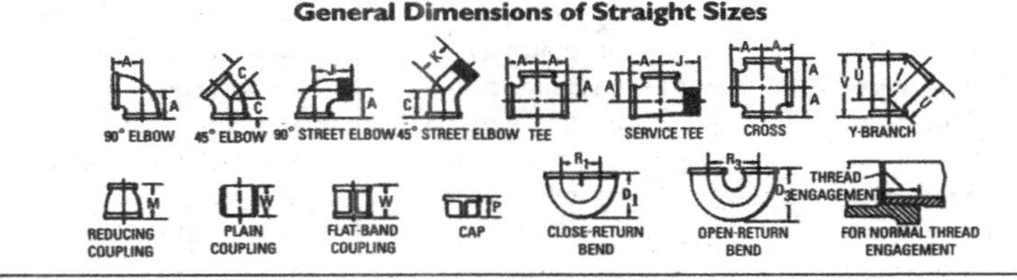

Size, in.	Dimensions, in.												
1/8	11/16	3/4	–	–	1	7/8	–	11/16	–	–	–	–	31/32
1/4	13/16	3/4	–	–	1 3/16	15/16	1 1/4	11/16	–	–	–	–	1 1/16
3/8	15/16	13/16	–	–	1 7/16	1 1/32	1 1/4	11/16	–	–	1 23/32	2 15/32	1 5/32
1/2	1 1/8	7/8	1 3/4	1 7/8	1 5/8	1 5/32	1 3/8	25/32	1 1/8	1 1/2	1 23/32	2 15/32	1 11/32
3/4	1 5/16	1	2 1/32	2 5/16	1 7/8	1 5/16	1 7/16	1 1/32	1 3/8	2	2 1/16	2 7/8	1 1/2
1	1 1/2	1 1/8	2 5/16	2 25/32	2 1/8	1 15/32	1 11/16	1 7/32	1 3/4	2 1/2	2 7/16	3 3/8	1 11/16
1 1/4	1 3/4	1 5/16	2 13/16	3 5/16	2 7/16	1 23/32	2 1/16	1 9/32	2 1/8	3	2 15/16	4 3/32	1 15/16
1 1/2	1 15/16	1 7/16	3 3/16	3 11/16	2 11/16	1 7/8	2 5/16	1 11/32	2 1/2	3 1/2	3 9/32	4 17/32	2 5/32
2	2 1/4	1 11/16	3 7/8	4 1/4	3 1/4	7 7/32	2 5/8	1 7/16	2 3/4	4	4 1/32	5 17/32	2 17/32
2 1/2	2 11/16	1 15/16	–	5 5/32	4	–	3 1/16	2	–	4 1/2	4 3/4	6 1/2	2 7/8
3	3 1/16	2 3/16	–	5 13/16	4 11/16	–	3 9/16	2 1/8	–	5	5 11/16	7 3/4	3 1/2
3 1/2	3 27/32	2 19/32	–	–	–	–	4	2 3/16	–	–	–	–	–
4	3 13/16	2 5/8	–	6 15/16	5 11/16	–	4 3/8	2 5/16	–	6	6 15/16	9	4 5/16
5	5 1/2	3 1/16	–	6 15/16	–	–	5 1/8	2 17/32	–	–	–	–	–
6	5 1/8	3 15/32	–	–	–	–	5 7/8	2 11/16	–	–	–	–	–

General dimensions of Crane standard malleable iron screwed fittings.

Standard Lengths of Nipples

	Standard Black, Right Hand		
Size, in.		**Kind of Nipples**	
1/8 to 1/2	Close, short, then by	1/2-inch lengths from 2	inches long to 6 inches long
	then by	1-inch lengths from 6	inches long to 12 inches long
3/4 and 1	Close, short, then by	1/2-inch lengths from 2 1/2	inches long to 6 inches long
	then by	1-inch lengths from 6	inches long to 12 inches long
1 1/4 to 2	Close, short, then by	1/2-inch lengths from 3	inches long to 6 inches long
	then by	1-inch lengths from 6	inches long to 12 inches long
2 1/2 and 3	Close, short, then by	1/2-inch lengths from 3 1/2	inches long to 6 inches long
	then by	1-inch lengths from 6	inches long to 12 inches long
3 1/2 and 4	Close, short, then by	1/2-inch lengths from 4 1/2	inches long to 6 inches long
	then by	1-inch lengths from 6	inches long to 12 inches long
5 and 6	Close, short, then by	1/2-inch lengths from 5	inches long to 6 inches long
	then by	1-inch lengths from 6	inches long to 12 inches long
8	Close, short, then by	1/2-inch lengths from 5 1/2	inches long to 6 inches long
	then by	1-inch lengths from 6	inches long to 12 inches long
10 and 12	Close, short, then by	1-inch lengths from 8	inches long to 12 inches long

Standard Galvanized, Right Hand
Up to and including 8-inch size, same lengths as black, right hand
Standard Black, Right and Left Hand
Up to and including 4-inch size, same lengths as black, right hand
Extra Strong Black, Right Hand
Up to and including 2-inch size, same lengths as standard, black, right hand

Length of Pipe for Fittings to Be Added to Actual Length of Run in Order to Obtain Equivalent Length

Size of Pipe, in.	Length to Be Added to Run, Ft				
	Standard Elbow	*Side-Outlet Tee*	*Gate Valve**	*Globe Valve**	*Valve**
½	1.3	3	0.3	14	7
¾	1.8	4	0.4	18	10
1	2.2	5	0.5	23	12
1¼	3.0	6	0.6	29	15
1½	3.5	7	0.8	34	18
2	4.3	8	1.0	46	22
2½	5.0	11	1.1	54	27
3	6.5	13	1.4	66	34
3½	8	15	1.6	80	40
4	9	18	1.9	92	45
5	11	22	2.2	112	56
6	13	27	2.8	136	67
8	17	35	3.7	180	92
10	21	45	4.6	230	112
12	27	53	5.5	270	132
14	30	63	6.4	310	152

*Valve in full open position.

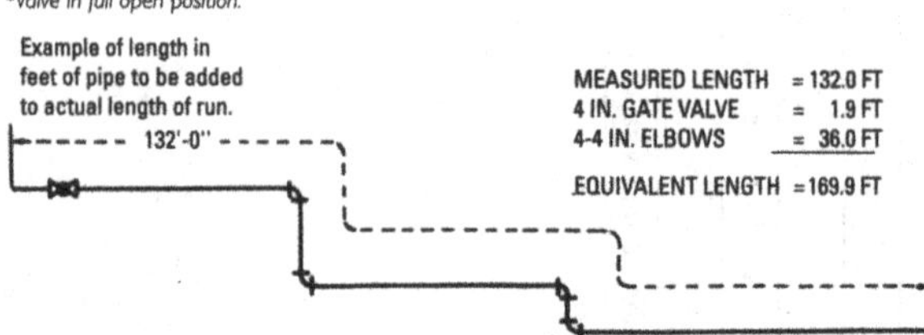

Equivalent Length of New Straight Pipe for Valves and Fittings for Turbulent Flow

Fittings			Pipe Size																				
			1/4	3/8	1/2	3/4	1	1 1/4	1 1/2	2	2 1/2	3	4	5	6	8	10	12	14	16	18	20	24
Regular 90° Ell	Screwed	Steel	2.3	3.1	3.6	4.4	5.2	6.6	7.4	8.5	9.3	11	13	—	—	—	—	—	—	—	—	—	—
		C. I.	—	—	—	—	—	—	—	—	—	9	11	—	—	—	—	—	—	—	—	—	—
	Flanged	Steel	—	—	0.92	1.2	1.6	2.1	2.4	3.1	3.6	4.4	5.9	7.3	8.9	12	14	17	18	21	23	25	30
		C. I.	—	—	—	—	—	—	—	—	—	3.6	4.8	—	7.2	9.8	12	15	17	19	22	24	28
Long Radius 90° Ell	Screwed	Steel	1.5	2	2.2	2.3	2.7	3.2	3.4	3.6	3.6	4	4.6	—	—	—	—	—	—	—	—	—	—
		C. I.	—	—	—	—	—	—	—	—	—	3.3	3.7	—	—	—	—	—	—	—	—	—	—
	Flanged	Steel	—	—	1.1	1.3	1.6	2	2.3	2.7	2.9	3.4	4.2	5	5.7	7	8	9	9.4	10	11	12	14
		C. I.	—	—	—	—	—	—	—	—	—	2.8	3.4	—	4.7	5.7	6.8	7.8	8.6	9.6	11	11	13
Regular 45° Ell	Screwed	Steel	0.34	0.52	0.71	0.92	1.3	1.7	2.1	2.7	3.2	4	5.5	—	—	—	—	—	—	—	—	—	—
		C. I.	—	—	—	—	—	—	—	—	—	3.3	4.5	—	—	—	—	—	—	—	—	—	—
	Flanged	Steel	—	—	0.45	0.59	0.81	1.1	1.3	1.7	2	2.6	3.5	4.5	5.6	7.7	9	11	13	15	16	18	22
		C. I.	—	—	—	—	—	—	—	—	—	2.1	2.9	—	4.5	6.3	8.1	9.7	12	13	15	17	20
Tee-Line Flow	Screwed	Steel	0.79	1.2	1.7	2.4	3.2	4.6	5.6	7.7	9.3	12	17	—	—	—	—	—	—	—	—	—	—
		C. I.	—	—	—	—	—	—	—	—	—	9.9	14	—	—	—	—	—	—	—	—	—	—
	Flanged	Steel	—	—	0.69	0.82	1	1.3	1.5	1.8	1.9	2.2	2.8	3.3	3.8	4.7	5.2	6	6.4	7.2	7.6	8.2	9.6
		C. I.	—	—	—	—	—	—	—	—	—	1.9	2.2	—	3.1	3.9	4.6	5.2	5.9	6.5	7.2	7.7	8.8
	Screwed	Steel	2.4	3.5	4.2	5.3	6.6	8.7	9.9	12	13	17	21	—	—	—	—	—	—	—	—	—	—
		C. I.	—	—	—	—	—	—	—	—	—	14	17	—	—	—	—	—	—	—	—	—	—

(continued)

(continued)

Fittings			Pipe Size																				
			1/4	3/8	1/2	3/4	1	1 1/4	1 1/2	2	2 1/2	3	4	5	6	8	10	12	14	16	18	20	24
Tee-Branch Flow	Flanged	Steel	—	—	2	2.6	3.3	4.4	5.2	6.6	7.5	9.4	12	15	18	24	30	34	37	43	47	52	62
		C. I.	—	—	—	—	—	—	—	—	—	7.7	10	—	15	20	25	30	35	39	44	49	57
180° Return Bend	Screwed	Steel	2.3	3.1	3.6	4.4	5.2	6.6	7.4	8.5	9.3	11	13	—	—	—	—	—	—	—	—	—	—
		C. I.	—	—	—	—	—	—	—	—	—	9	11	—	—	—	—	—	—	—	—	—	—
	Reg. Flanged	Steel	—	—	0.92	1.2	1.6	2.1	2.4	3.1	3.6	4.4	5.9	7.3	8.9	12	14	17	18	21	23	25	30
		C. I.	—	—	—	—	—	—	—	—	—	3.6	4.8	—	7.2	9.8	12	15	17	19	22	24	28
	Long Rad. Flanged	Steel	—	—	1.1	1.3	1.6	2	2.3	2.7	2.9	3.4	4.2	5	5.7	7	8	9	9.4	10	11	12	14
		C. I.	—	—	—	—	—	—	—	—	—	2.8	3.4	—	4.7	5.7	6.8	7.8	8.6	9.6	11	11	13
Globe Valve	Screwed	Steel	21	22	22	24	29	37	42	54	62	79	110	—	—	—	—	—	—	—	—	—	—
		C. I.	—	—	—	—	—	—	—	—	—	65	86	—	—	—	—	—	—	—	—	—	—
	Flanged	Steel	—	—	38	40	45	54	59	70	77	94	120	150	190	260	310	390	—	—	—	—	—
		C. I.	—	—	—	—	—	—	—	—	—	77	99	—	150	210	270	330	—	—	—	—	—
Gate Valve	Screwed	Steel	0.32	0.45	0.56	0.67	0.84	1.1	1.2	1.5	1.7	1.9	2.5	—	—	—	—	—	—	—	—	—	—
		C. I.	—	—	—	—	—	—	—	—	—	1.6	2	—	—	—	—	—	—	—	—	—	—
	Flanged	Steel	—	—	—	—	—	—	—	2.6	2.7	2.8	2.9	3.1	3.2	3.2	3.2	3.2	3.2	3.2	3.2	3.2	3.2
		C. I.	—	—	—	—	—	—	—	—	—	2.3	2.4	—	2.6	2.7	2.8	2.9	3	3	3	3	3
Angle Valve	Screwed	Steel	12.8	15	15	15	17	18	18	18	18	18	18	—	—	—	—	—	—	—	—	—	—
		C. I.	—	—	—	—	—	—	—	—	—	15	15	—	—	—	—	—	—	—	—	—	—
	Flanged	Steel	—	—	15	15	17	18	18	21	22	28	38	50	63	90	120	140	160	190	210	240	300
		C. I.	—	—	—	—	—	—	—	—	—	23	31	—	52	74	98	120	150	170	200	230	280

Fitting	Connection	Material																					
Swing Check Valve	Screwed	Steel	7.2	7.3	8	8.8	11	13	15	19	22	27	38	—	—	—	—	—	—	—	—	—	—
		C. I.	—	—	—	—	—	—	—	—	—	22	31	—	—	—	—	—	—	—	—	—	—
	Flanged	Steel	—	—	3.8	5.3	7.2	10	12	17	21	27	38	50	63	90	120	140	—	—	—	—	—
		C. I.	—	—	—	—	—	—	—	—	—	22	31	—	52	74	98	120	—	—	—	—	—
Coupling or Union	Screwed	Steel	0.14	0.18	0.21	0.24	0.29	0.36	0.39	0.45	0.47	0.53	0.65	—	—	—	—	—	—	—	—	—	—
		C. I.																					
Bell Mouth Inlet		Steel	—	—	—	—	—	—	—	—	—	0.44	0.62	—	—	—	—	—	—	—	—	—	—
		C. I.	0.04	0.07	0.1	0.13	0.18	0.26	0.31	0.43	0.52	0.67	0.95	1.3	1.6	2.3	2.9	3.5	4	4.7	5.3	6.1	7.6
Square Mouth Inlet		Steel	—	—	—	—	—	—	—	—	—	0.55	0.77	—	1.3	1.9	2.4	3	3.6	4.3	5	5.7	7
		C. I.	0.44	0.68	0.96	1.3	1.8	2.6	3.1	4.3	5.2	6.7	9.5	13	16	23	29	35	40	47	53	61	76
Reentrant Pipe		Steel	—	—	—	—	—	—	—	—	—	5.5	7.7	—	13	19	24	30	36	43	50	57	70
		C. I.	0.88	1.4	1.9	2.6	3.6	5.1	6.2	8.5	10	13	19	25	32	45	58	70	80	95	110	120	150
			—	—	—	—	—	—	—	—	—	11	15	—	26	37	49	61	73	86	100	110	140
Y-Strainer			—	4.6	5	6.6	7.7	18	20	27	29	34	42	53	61								

Sudden Enlargement (V_1 → V_2)

$$h = \frac{(V_1 - V_2)^2}{(2g)} \text{ feet of liquid; If } V_2 = 0 \; h = \frac{V^2}{(2g)} \text{ feet of liquid}$$

Courtesy The Hydraulic Institute (reprinted from the Standards of the Hydraulic Institute, Eleventh Edition, Copyright 1965).

Diagram Showing Resistance of Valves and Fittings to the Flow of Liquids

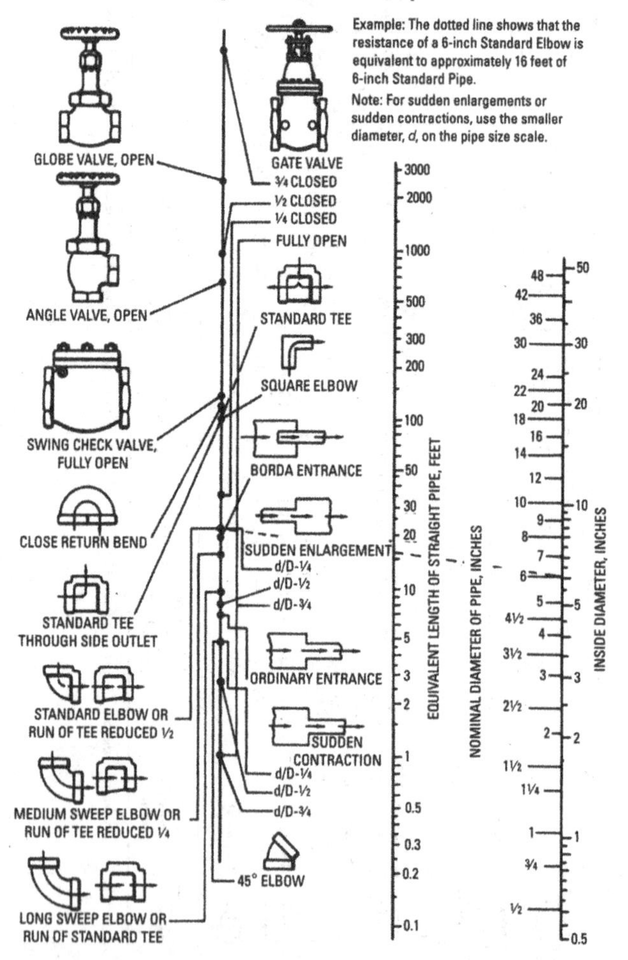

Flanged Pipe Connections

Flanged pipe connections are widely used, particularly on larger pipes, as they provide a practical and economical piping connection system. Flanges are commonly connected to the pipe by screw threads or by welding. Several types of flange facings are in use, the simplest of which are the plain *flat face* and the *raised face.*

The plain flat-faced flange is usually used for cast-iron flanges where pressures are less than 125 pounds. Higher-pressure cast-iron flanges and steel flanges are made with a raised face. Generally, full-face gaskets are used with flat-face flanges and ring gaskets with raised-face flanges. The function of the gasket is to provide a loose, compressible substance between the faces with sufficient body resiliency and strength to make the flange connection leakproof.

Standard Cast-Iron Companion Flanges and Bolts (for working pressures up to 125 lb/in.^{2} steam, 175 lb/in.^{2} WOG)

Size, in.	Diameter of Flange, in.	Bolt Circle, in.	Number of Bolts	Size of Bolts, in.	Length of Bolts, in.
3/4	3 1/2	2 1/2	4	3/8	1 3/8
1	4 1/4	3 1/8	4	1/2	1 1/2
1 1/4	4 5/8	3 1/2	4	1/2	1 1/2
1 1/2	5	3 7/8	4	1/2	1 3/4
2	6	4 3/4	4	5/8	2
2 1/2	7	5 1/2	4	5/8	2 1/2
3	7 1/2	6	4	5/8	2 1/2
3 1/2	8 1/2	7	8	5/8	2 1/2
4	9	7 1/2	8	5/8	2 3/4
5	10	8 1/2	8	3/4	3
6	11	9 1/2	8	3/4	3
8	13 1/2	11 3/4	8	3/4	3 1/4
10	16	14 1/4	12	7/8	3 1/2
12	19	17	12	7/8	3 3/4
14	21	18 3/4	12	1	4 1/4
16	23 1/2	21 1/4	16	1	4 1/4

Extra-Heavy Cast-Iron Companion Flanges and Bolts (for working pressure up to 250 lb/in.² steam, 400 lb/in.² WOG)

Pipe Size, in.	*Diameter of Flanges*	*Diameter of Bolt Circle*	*Number of Bolts*	*Diameter of Bolts*	*Length of Bolts*
1	4⅞	3½	4	⅝	2¼
1¼	5¼	3⅞	4	⅝	2½
1½	6⅛	4½	4	¾	2½
2	6½	5	8	⅝	2½
2½	7½	5⅞	8	¾	3
3	8¼	6⅝	8	¾	3¼
3½	9	7¼	8	¾	3¼
4	10	7⅞	8	¾	3½
5	11	9¼	8	¾	3¾
6	12½	10⅝	12	¾	3¾
8	15	13	12	⅞	4¼
10	17½	15¼	16	1	5
11	20½	17¾	16	1⅛	5½
14 OD	23	20¼	20	1⅛	5¾
16 OD	25½	22½	20	1¼	6

Flanged Bolting

The assembly and tightening of a pipe-flange connection is a relatively simple operation; however, certain practices must be followed to obtain a leakproof connection. The gasket must line up evenly with the inside bore of the flange face, with no portion of it extending into the bore. When tightening the bolts, the flange faces must be kept parallel and the bolts must be tightened uniformly.

The tightening sequence for round flanges is shown in part A of the figure. The sequence is to lightly tighten the first bolt, then move directly across the circle for the second bolt, then move a quarter way around the circle for the third, and directly across for the fourth, continuing the sequence until all are tightened.

When tightening an oval flange, the bolts are tightened across the short centerline, as shown in part B of the figure.

A four-bolt flange, either round or square, is tightened with a simple criss-cross sequence, as shown in part C of the figure.

Do not snug up the bolts on the first go-around. This can tilt the flanges out of parallel. If using an impact wrench, set the wrench at about one-half final torque for the first go-around. Pay particular attention to the hard-to-reach bolts.

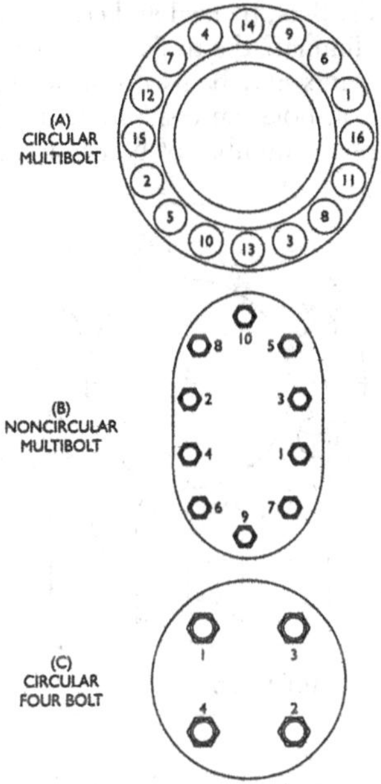

(A) CIRCULAR MULTIBOLT

(B) NONCIRCULAR MULTIBOLT

(C) CIRCULAR FOUR BOLT

Tightening sequences for round flanges.

Pipe Flange Bolt-Hole Layout

Mating pipe flanges to other flanges or circular parts requires correct layout of bolt holes. In addition, the holes must be located around the circle to line up when the flanges are mated. The usual practice is to specify the location of the holes as either "on" or "off" the vertical centerline. The shop term commonly applied to "on" the centerline layout is *one hole up* and to "off" the centerline is two holes up. An "off" the centerline or two holes up layout is illustrated in the figure.

Although bolt holes may be laid out with a protractor using angular measurements to obtain uniform spacing, this method is most satisfactory when there are six or fewer holes. Also, layout by stepping off spacing around the circle with dividers, by trial and error, is a time-consuming operation. To eliminate the trials and errors, a system of multipliers or constants may be used to calculate the chordal distance between bolt-hole centers. Simply multiply the constant for the appropriate number of bolt holes, as shown in the following table by the bolt circle diameter to determine the chordal distance between holes.

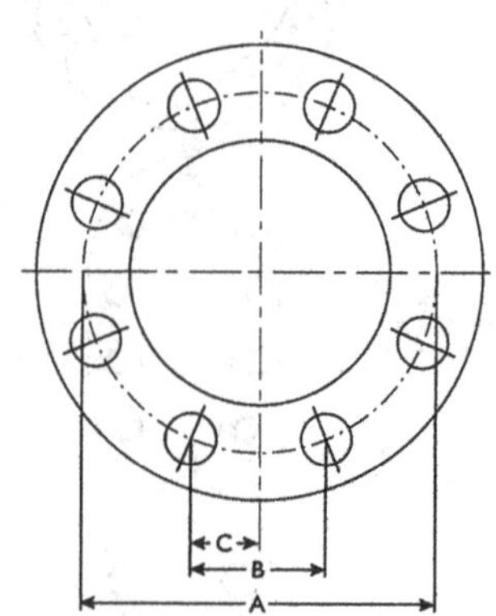

Bolt-hole layout.

Flange Hole Constants

No. of Bolt Holes	Constant
4	0.707
6	0.500
8	0.383
10	0.309
12	0.259
16	0.195
20	0.156
24	0.131
28	0.112
32	0.098
36	0.087
40	0.079

Layout Procedure

As shown in the preceding figure:

1. Lay out horizontal and vertical centerlines.
2. Lay out bolt circle.
3. Find value of B (multiply bolt circle diameter by constant.
4. For two holes up layout, divide B by 2 for value of C.
5. Measure distance C off the centerline and locate the center of the first bolt hole.
6. Set dividers to dimension B and layout center points by swinging arcs, starting from first center point.

Using Pipe Sealant

Using a lubricant called *dope* allows thread surfaces to deform and mate without galling and seizing. The dope also helps plug openings resulting from improper threads, and it acts as a cement.

The requirements for tight makeup of threaded pipe connections are good-quality threads, clean threads, proper dope for the application, and slow final makeup to avoid heat generation.

A recent addition to the sealant field is the use of TFE pipe tape. This material is wound onto the threads in a direction opposite to the lead of the thread and forms a good seal and acts as a lubricant as the threads are pulled up. Take care not to overlap the TFE tape at the end of the pipe threads, because this might cause parts of the tape to enter the final fluid stream after the line is commissioned. This can cause problems if the tape becomes lodged in a critical sensing or flow-control device used in the line.

Calculating Offsets

In pipe fitting, an *offset* is a change of direction (other than 90°) in a pipe bringing one part out of (but parallel with) the line of another.

45-Degree Offsets

The *travel* distance of a 45° offset is calculated in the same manner as the diagonal of a square. Multiply the distance across flats by 1.414. The *run* and *offset* represent the two equal sides of a square, and the *travel*, the diagonal, is shown in the figure.

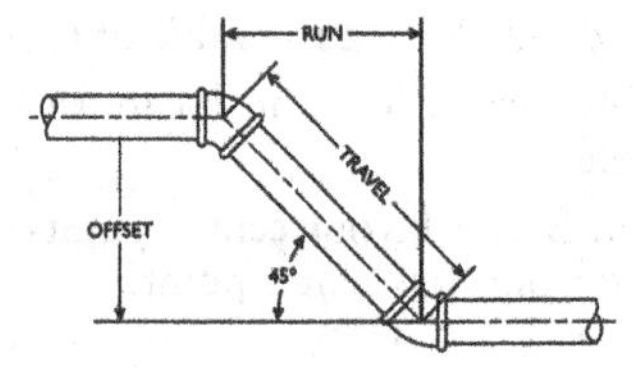

Diagonal of an offset, illustrating run, offset, and travel.

There are also occasions when the travel is known and the offset and run dimensions are wanted. This is calculated in the same manner as finding the distance across the flats of a square when the across-corners dimension is known. Multiply the travel dimension by 0.707.

Examples

What is the travel of a 16-inch, 45° offset?

$$16 \times 1.414 = 22.625 \text{ or } 22\tfrac{5}{8} \text{ inches}$$

What are the offset and run of a 45° offset having a travel of 26 inches?

$$26 \times 0.707 = 18.382 \text{ or } 18\tfrac{3}{8} \text{ inches}$$

Other Offsets

The dimensions of piping offsets for several other common angles may be calculated by multiplying the known values by the appropriate constants.

Common Angle Pipe Offsets—Multipliers

Angle	*To Find Travel, Offset Known*	*To Find Travel, Run Known*	*To Find Run, Travel Known*	*To Find Offset, Travel Known*
60°	1.155	2.000	0.500	0.866
30°	2.000	1.155	0.866	0.500
22½°	2.613	1.082	0.924	0.383
11¼°	5.126	1.000	0.980	0.195

45-Degree Rolling Offsets

The 45° offset is often used to offset a pipeline in a plane other than the horizontal or vertical. This is done by rotating the offset out of the horizontal or vertical plane, and it is known as a *rolling offset*. The rolling offset can best be visualized as contained in an imaginary isometric box, as shown in the figure.

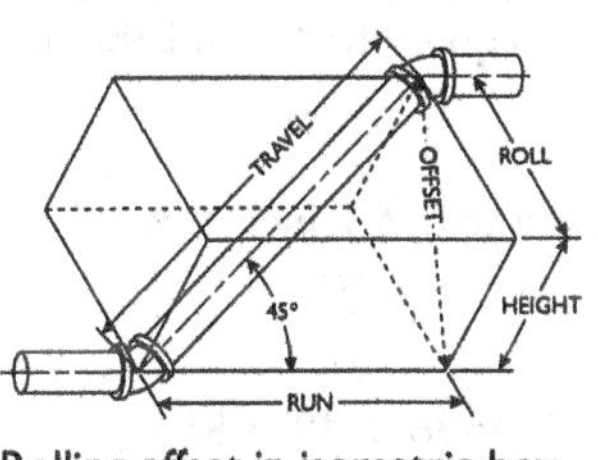

Rolling offset in isometric box.

The run and offset distances are equal, as they are in the plain 45° offset; however, there are two additional dimensions, *roll* and *height*. Two right-angle triangles must now be considered. The original one remains the same, with the offset and run as equal sides and the travel as the hypotenuse. The new triangle has the roll and height as sides and the offset as the hypotenuse.

The method of finding distances for plain 45° offsets is also used for calculating rolling offset distances. The sum-of-the-squares equations are used to find the values of the second triangle.

Sum of the Squares

The sum-of-the-squares equation states that the hypotenuse of a right-angle triangle squared is equal to the side opposite squared plus the side adjacent squared. This equation is commonly written as follows:

$$c^2 = a^2 + b^2$$

Substitution of pipe-offset terms in this equation and its rearrangements gives the following equations:

$$\text{offset}^2 = \text{roll}^2 + \text{height}^2$$

$$\text{run}^2 = \text{roll}^2 + \text{height}^2$$

$$\text{roll}^2 = \text{offset}^2 - \text{height}^2$$

$$\text{roll}^2 = \text{run}^2 - \text{height}^2$$

$$\text{height}^2 = \text{offset}^2 - \text{roll}^2$$

$$\text{height}^2 = \text{run}^2 - \text{roll}^2$$

Depending on what the known values are, it may sometimes be necessary to solve two equations to find the distance wanted.

Examples

What is the *travel* for a 6-inch *roll* with a 7-inch *height*?

Using the formula:

$$\text{offset}^2 = \text{roll}^2 + \text{height}^2$$

$$\text{offset}^2 = (6 \times 6) + (7 \times 7) = 36 + 49 = 85$$

$$\text{offset} = \sqrt{85} = 9.22 = 9\tfrac{7}{32} \text{ inches}$$

Followed by:

$$\text{travel} = \text{offset} \times 1.414$$

$$\text{travel} = 9\tfrac{7}{32} \times 1.414 = 13.035 = 13\tfrac{1}{32} \text{ inches}$$

What is the *roll* for an 11-inch *offset* with an 8-inch *height*? Using the formula:

$$\text{roll}^2 = \text{offset}^2 - \text{height}^2$$

$$\text{roll}^2 = (11 \times 11) - (8 \times 8) = 121 - 64 = 57$$

$$\text{roll} = \sqrt{57} = 7.55 = 7\tfrac{9}{16} \text{ inches}$$

What is the *height* for a 16-inch *offset* with a 12-inch *roll*? Using the formula:

$$\text{height}^2 = \text{offset}^2 - \text{roll}^2$$

$$\text{height}^2 = (16 \times 16) - (12 \times 12) = 256 - 144 = 112$$

$$\text{height} = \sqrt{122} = 10.583 = 10\tfrac{9}{32} \text{ inches}$$

Valves

Valve Drafting Symbols

	Flanged	*Screwed*	*Bell and Spigot*	*Welded*	*Soldered*
Angle Valve					
Check (also angle check)					
Gate (also angle gate) (Elevation)					
Gate (also angle gate) (Plan)					

	Flanged	*Screwed*	*Bell and Spigot*	*Welded*	*Soldered*
Globe (also angle globe) (Elevation)					
Globe (Plan)					
Automatic Valve					
By-pass					
Governor-operated					
Reducing					
Check valve (straight way)					
Cock					
Diaphragm valve					

(continued)

(continued)

	Flanged	***Screwed***	***Bell and Spigot***	***Welded***	***Soldered***
Float valve					
Gate valve*					
Motor-operated					
Globe valve					
Hose valve (also hose globe); Angle (also hose angle)	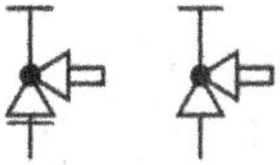				
Gate	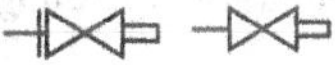				
Globe					
lockshield valve					
Quick-opening valve					
Safety valve					

**Also used for General Stop Valve symbol when amplified by specification.*

Symbols courtesy of Mechanical Contractors Association of America, Inc.

Valve Operating Ranges

Type of Valve	Sizes, in. Minimum	Sizes, in. Maximum	Sizes, mm Minimum	Sizes, mm Maximum
Gate valve	⅛	48	3	1220
Globe valve	⅛	30	3	760
Butterfly valve	2	72	50	1830
Butterfly-neck valve	1	72	1	1830
Ball valve	¼	48	6	1220
Swing-check valve	¼	24	6	610
Y-Type swing check valve	¼	6	6	150
Lift-check valve	¼	10	6	250

Valve Operating Temperatures

Type of Valve	Operating Temperature Range Minimum, °F	Operating Temperature Range Maximum, °F	Operating Temperature Range Minimum, °C	Operating Temperature Range Maximum, °C
Gate valve	–455	1250	–272	675
Globe valve	–455	1000	–272	540
Butterfly valve	–20	1000	–30	538
Butterfly-neck valve	0	500	–18	260
Ball valve	–65	575	–55	300
Swing-check valve	0	1200	–18	540
Y-Type swing-check valve	0	1200	–18	540
Lift-check valve	0	1200	–18	540

Valve Problems

Valve manufacturers usually provide information for servicing and repairing their valves. When the information is not available, as is the case on older heating and cooling systems, troubleshooting and correcting the valve problem will depend on experience or educated guesswork. Experience has shown that most valve problems can be traced to one of the following sources:

- Stuffing-box leakage
- Seat leakage
- Damaged stem

Valve Stuffing-Box Leakage

Leakage around the valve stuffing box is usually an indication that the stuffing must be adjusted or replaced. This leakage does not occur when the valve is completely opened or closed. Therefore, an absence of leakage is not necessarily an indication that the valve is functioning normally.

Once you have detected leakage, first check whether adjusting the packing will stop it. If it is a bolted bonnet valve, turn the packing gland nuts (or gland stud nuts) clockwise alternately with no more than ¼ turn on each until leakage stops. If you are dealing with a screwed and union-bonnet valve, turn the packing nut clockwise until the leakage stops. If the leakage will not respond to adjustment, the packing must be replaced.

The procedure for replacing the packing in most valves may be summarized as follows:

1. Remove the handwheel nut and the handwheel.
2. Remove the packing nut.
3. Slip the packing gland off of the stem.
4. Replace the packing.
5. Reassemble in reverse order.

The procedure used with bolted-bonnet outside screw and yoke valves is a little more complicated. On Y-valves of this type, it is necessary to remove the gland flange and gland follower before replacing the packing. On globe and angle valves, the stud nuts and upper valve assembly must be removed.

Because of their design and construction, the problem of stuffing-box leakage does not occur with check valves.

Valve Seat Leakage

Leakage of water from the valve body is usually an indication that the wedge, disc, or seat ring needs replacing. For most valves, the procedure for doing this may be summarized as follows:

1. Open the valve.
2. Remove the bonnet and other components of the upper valve assembly.
3. Run the stem down by turning it in a clockwise direction.
4. Remove the wedge or disc from the stem and replace if necessary.
5. Remove the seat ring with a seat-ring wrench and replace if necessary.
6. Reassemble in reverse order.

The disc, disc assembly, and ball are all possible sources of seat leakage in check valves. Access these components by removing the valve cap (counterclockwise), side plugs, and pins. Reassemble in reverse order.

Valve Stem Damage

Sometimes the threads on valve stems become worn or damaged, making the valves inoperable. When this occurs, the stems must be replaced. Before the stem can be replaced, however, all pressure must be removed. Then, with pressure removed, disconnect and remove the bonnet and upper valve assembly. The remainder of the procedure depends on the type of stem used (that is, rising or nonrising stem) and other design factors. The procedure may be summarized as follows:

1. Run the stem down by turning it clockwise.
2. Rotate the stem clockwise until the stem threads are completely out of the threaded portion of the upper bushing.
3. Pull the stem out of the stuffing box.
4. Remove the wedge or disc from the stem.
5. Replace the old stem with a new one.
6. Reassemble in reverse order with new packing and gasket (when applicable).

Gate, Globe, Angle, Butterfly, and Check Valve Details

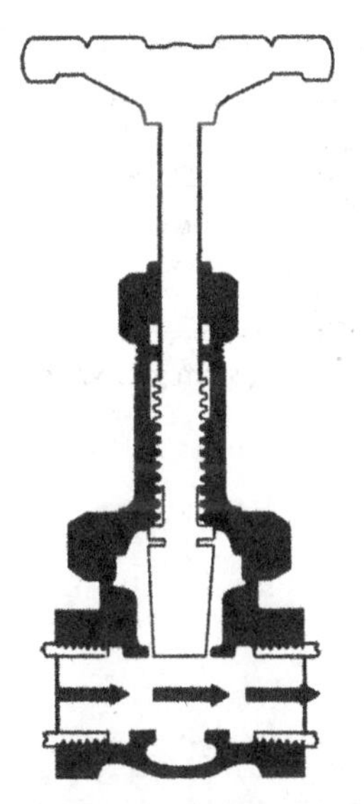

Gate valve flow characteristics.
(Courtesy Wm. Powell Co.)

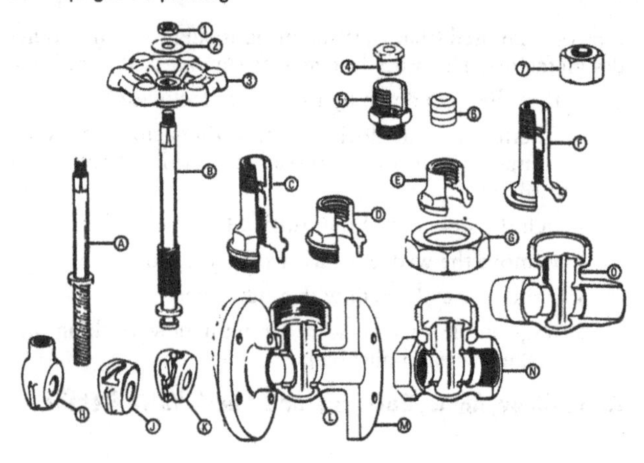

1. HANDWHEEL NUT
2. IDENTIFICATION PLATE
3. HANDWHEEL
4. PACKING GLAND
5. PACKING-BOX SPUD (NONRISING STEM VALVES ONLY)
6. PACKING
7. PACKING NUT

A. STEM NONRISING STEM VALVE
B. STEM RISING STEM VALVES
C. SCREWED-IN BONNET RISING STEM VALVE
D. SCREWED-IN BONNET NONRISING STEM VALVE
E. UNION BONNET NONRISING STEM VALVE
F. UNION BONNET RISING STEM VALVES
G. BONNET RING
H. SOLID WEDGE NONRISING STEM VALVES
J. SOLID WEDGE RISING STEM VALVE
K. DOUBLE WEDGE-RISING STEM VALVES
L. SEAT RING
M BODY-FLANGED ENDS
N. BODY-THREADED ENDS
O. BODY-SOLDER JOINT ENDS

Screwed and union bonnet rising and nonrising stem gate valves.
(Courtesy Wm. Powell Co.)

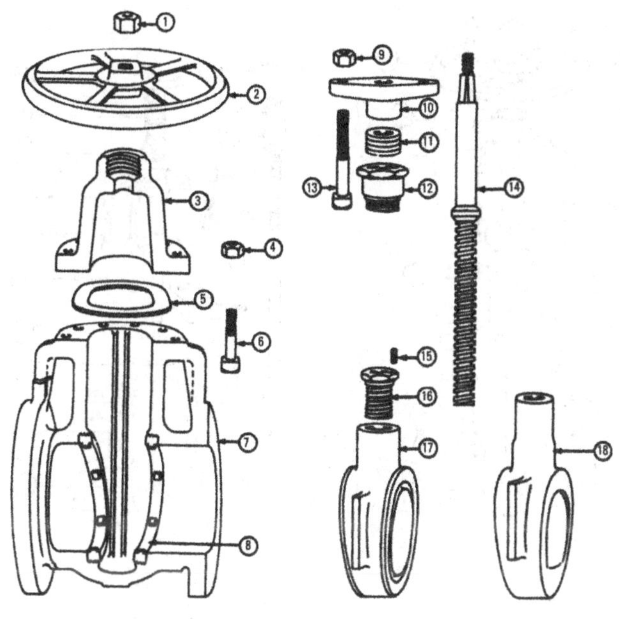

1. HANDWHEEL NUT
2. HANDWHEEL
3. BONNET
4. BODY NUTS
5. GASKET
6. BODY BOLTS
7. BODY
8. SEAT RING
9. GLAND NUTS
10. PACKING GLAND
11. PACKING
12. SPUD
13. GLAND BOLTS
14. STEM
15. WEDGE NUT SET SCREW
16. WEDGE NUT
17. WEDGE
18. WEDGE

Bolted-bonnet inside screw nonrising-stem gate valve.

(Courtesy Wm. Powell Co.)

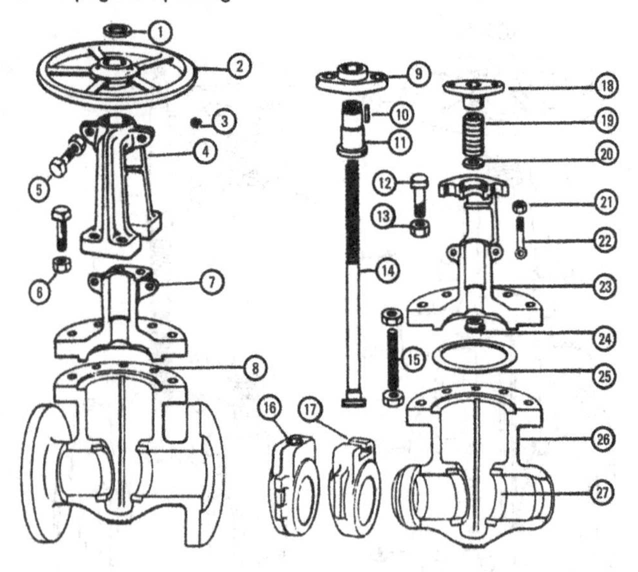

1. STEM BUSHING NUT
2. HANDWHEEL
3. LUBRICANT FITTING
4. YOKE-ARMS
5. YOKE-ARM EAR BOLT AND NUT
6. BONNET BOLT AND NUT
7. TWO-PIECE BONNET
8. BODY-FLANGED ENDS
9. BEARING CAP
10. HANDWHEEL KEY
11. STEM BUSHING
12. BEARING CAP BOLT
13. BEARING CAP NUT
14. STEM
15. BODY STUD AND NUTS
16. SPLIT WEDGE
17. SOLID WEDGE
18. PACKING GLAND
19. PACKING
20. PACKING WASHER
21. EYEBOLT NUT
22. EYEBOLT
23. ONE-PIECE BONNET
24. PACK-UNDER-PRESSURE BUSHING
25. GASKET
26. BODY-WELDED ENDS
27. SEAT RING

Bolted-bonnet outside screw and yoke rising-stem gate valve.

(Courtesy Wm. Powell Co.)

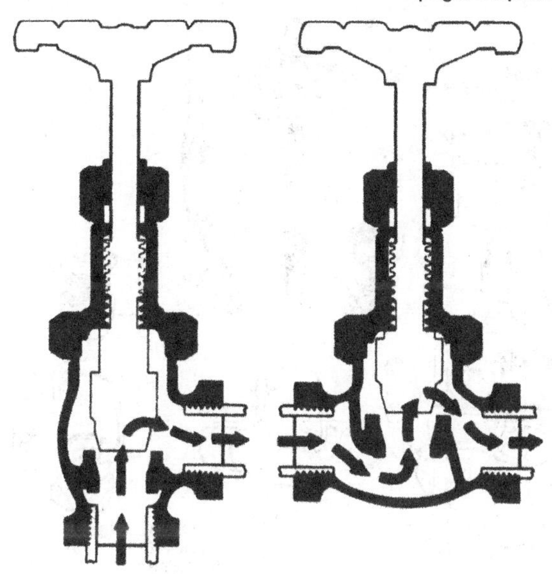

Sectional view of a globe valve and an angle valve.
(Courtesy Wm. Powell Co.)

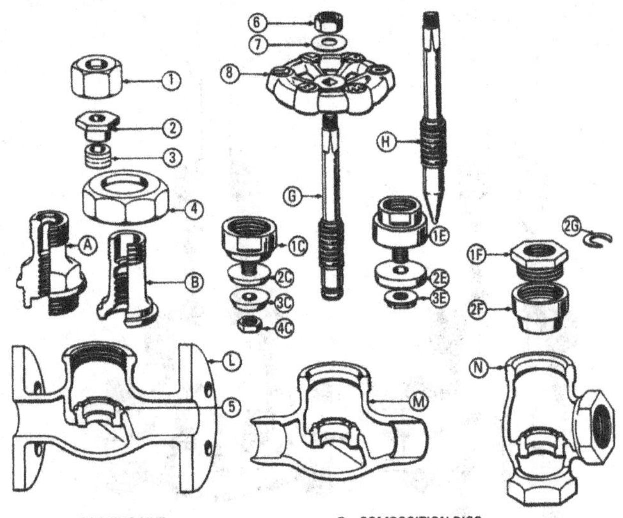

1. PACKING NUT
2. PACKING GLAND
3. PACKING
4. BONNET RING
5. SEAT RING
6. HANDWHEEL NUT
7. IDENTIFICATION PLATE
8. HANDWHEEL

A. SCREWED-IN BONNET
B. UNION BONNET
C. HI-LO DISC
    1C DISC HOLDER
    2C NONMETALLIC DISC
    3C DISC PLATE
    4C DISC NUT
E. COMPOSITION DISC
    1E DISC HOLDER
    2E NONMETALLIC DISC
    3E DISC-LOCKNUT WASHER
F. DISC LOCKNUT
    1F DISC NUT
    2F DISC
G. STEM-DISC LOCKNUT (HORSESHOE RING) TYPE
    2G HORSESHOE RING
H. STEM-NEEDLE DISC TYPE
L. BODY-GLOBE-FLANGED ENDS
M. BODY-GLOBE-SOLDER JOINT ENDS
N. BODY-ANGLE-THREADED ENDS

Screwed- and union-bonnet globe and angle valves.

(Courtesy Wm. Powell Co.)

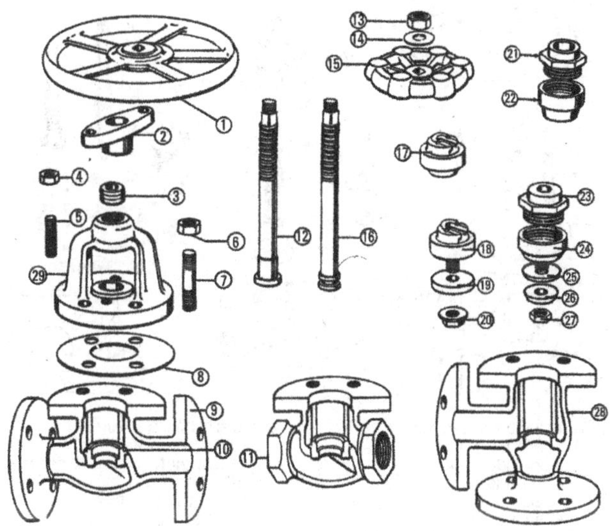

1. HANDWHEEL (ROUND)
2. PACKING GLAND
3. PACKING
4. GLAND STUD NUT
5. GLAND STUD
6. YOKE STUD NUT
7. YOKE STUD
8. GASKET
9. BODY-GLOBE-FLANGED ENDS
10. SEAT RING
11. BODY-GLOBE-THREADED ENDS
12. STEM-DISC LOCKNUT TYPE
13. HANDWHEEL NUT
14. IDENTIFICATION PLATE
15. HANDWHEEL
16. STEM-SLIP-ON TYPE
17. DISC-ONE-PIECE-SLIP-ON
18. COMPOSITION DISC HOLDER
19. COMPOSITION DISC
20. DISC NUT
21. DISC LOCKNUT
22. DISC
23. DISC LOCKNUT-HI-LO DISC
24. DISC HOLDER
25. NONMETALLIC DISC
26. DISC PLATE
27. DISC NUT
28. BODY-ANGLE-FLANGED ENDS
29. YOKE

Bolted-bonnet outside screw and yoke globe and angle valves.

(Courtesy Wm. Powell Co.)

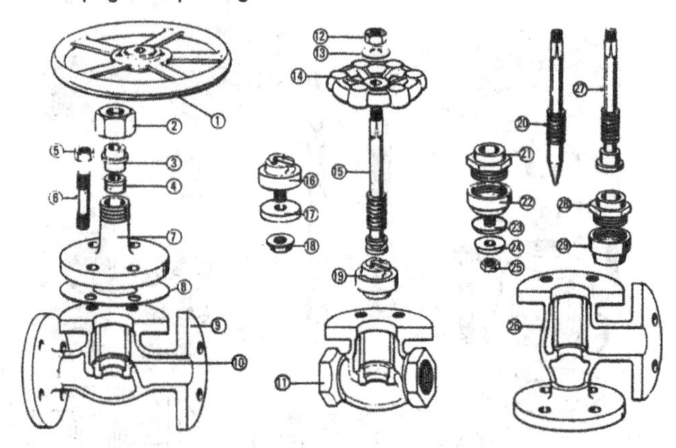

1. HANDWHEEL-ROUND
2. PACKING NUT
3. PACKING GLAND
4. PACKING
5. BODY NUT
6. BODY STUD
7. BONNER
8. GASKET
9. BODY-GLOBE-FLANGED ENDS
10. SEAT RING
11. BODY-GLOBE-THREADED ENDS
12. HANDWHEEL NUT
13. IDENTIFICATION PLATE
14. HANDWHEEL-NONHEATING
15. STEM-SLIP-ON DISC TYPE
16. COMPOSITION DISC HOLDER
17. NONMETALLIC DISC
18. DISC LOCKNUT
19. SLIP-ON DISC
20. STEM-NEEDLE DISC TYPE
21. HI-LO DISC LOCKNUT
22. DISC HOLDER
23. NON-METALLIC DISC
24. DISC PLATE
25. DISC NUT
26. BODY-ANGLE-FLANGED ENDS
27. STEM-DISC LOCKNUT TYPE
28. DISC LOCKNUT
29. DISC

Bolted-bonnet inside screw globe and angle valve.

(Courtesy Wm. Powell Co.)

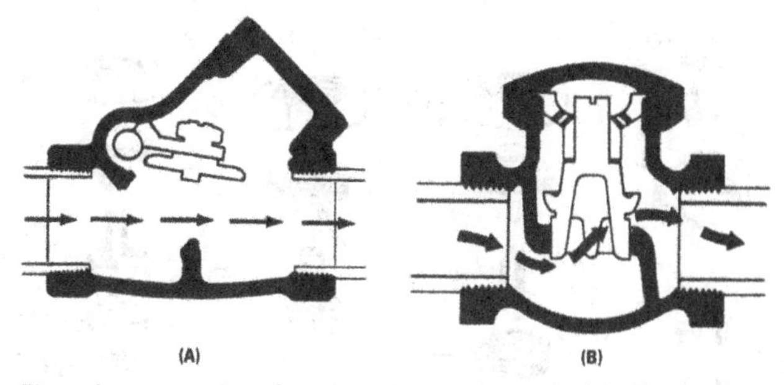

Flow characteristics of a swing-check valve and a lift-check valve.
(Courtesy Wm. Powell Co.)

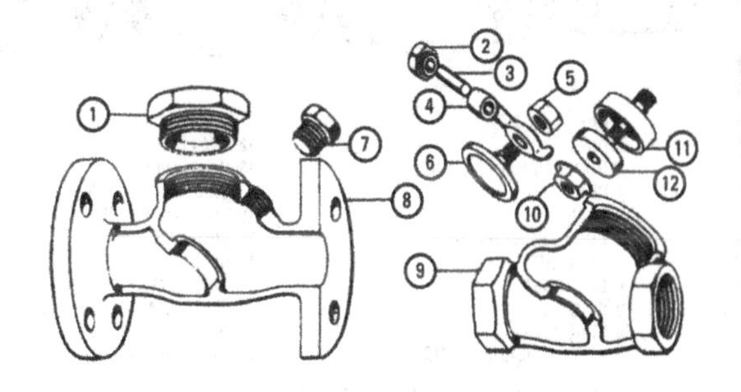

1. CAP
2. SLIDE PLUG
3. CARRIER PIN
4. CARRIER
5. DISC NUT
6. DISC
7. BUMPER PLUG
8. BODY-FLANGED END
9. BODY-THREADED END
10. DISC LOCKNUT-COMPOSITION DISC
11. DISC HOLDER
12. COMPOSITION DISC

Screwed-cap swing-check valve. *(Courtesy Wm. Powell Co.)*

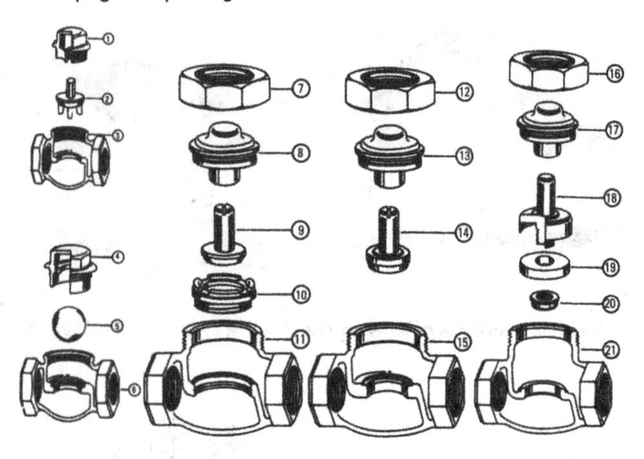

1. CAP
2. DISC
3. BODY-THREADED ENDS
4. CAP
5. BALL
6. BODY-THREADED ENDS
7. RING NUT
8. DISC GUIDE
9. DISC
10. SEAT RING
11. BODY-THREADED ENDS
12. RING NUT
13. DISC GUIDE
14. DISC
15. BODY-THREADED ENDS
16. RING NUT
17. DISC GUIDE
18. DISC HOLDER
19. DISC-COMPOSITION DISC
20. DISC NUT
21. BODY-THREADED ENDS

Screwed-cap horizontal-lift check value. *(Courtesy Wm. Powell Co.)*

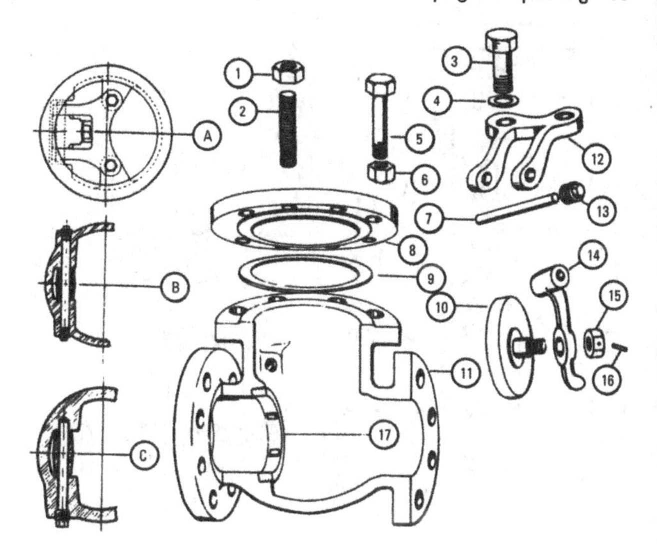

1. BODY NUT
2. BODY STUD
3. CAP SCREW
4. LOCK-WASHER
5. BODY BOLT
6. BODY BOLT NUT
7. DISC HOLDER PIN
8. CAP
9. GASKET
10. DISC
11. BODY-FLANGED ENDS
12. DISC HOLDER HANGER
13. PIPE PLUG
14. DISC HOLDER
15. DISC NUT
16. DISC NUT PIN
17. SEAT RING

A. DETAIL—HANGER TYPE DISC
B. DETAIL—PIN TYPE-TWO SIDE PLUGS
C. DETAIL—PIN TYPE-ONE SIDE PLUG

Bolted-cap swing-check valve. *(Courtesy Wm. Powell Co.)*

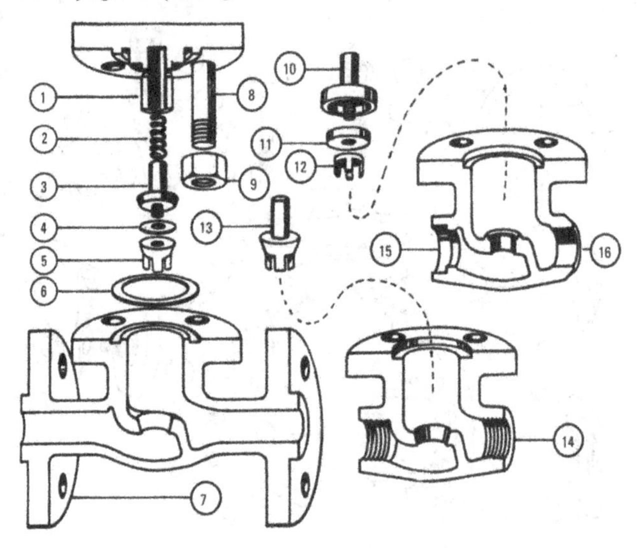

1. CAP—BOLTED
2. SPRING (OPTIONAL)
3. DISC HOLDER
4. TEFLON DISC
5. DISC GUIDE
6. GASKET
7. BODY-FLANGED ENDS
8. CAP BOLT
9. CAP BOLT NUT
10. COMPOSITION DISC HOLDER
11. NONMETALLIC DISC
12. DISC GUIDE
13. METAL DISC
14. BODY—THREADED ENDS
15. BODY—SILVER BRAZE ENDS
16. BODY—SOLDER JOINT ENDS

Bolted-cap horizontal-lift check valve. *(Courtesy Wm. Powell Co.)*

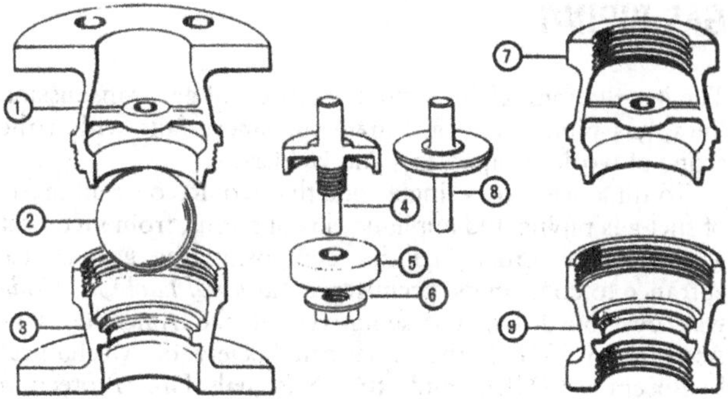

1. TOP BODY—FLANGED ENDS
2. BALL
3. BOTTOM BODY—FLANGED ENDS
4. DISC HOLDER
5. COMPOSITION DISC
6. DISC NUT
7. TOP BODY—THREADED ENDS
8. DISC
9. BOTTOM BODY—THREADED ENDS

Vertical check valve. *(Courtesy Wm. Powell Co.)*

GAS PIPING

The installation, testing, and connection of gas transmission lines, gas piping, and gas-fired appliances falls within the scope of work of pipefitters and welders.

To fill a need for a single code that would cover all facets of fuel-gas piping and appliance installations, from meter set assemblies or other facilities composing the gas service entrance to consumers' premises, a *National Fuel Gas Code* was developed by representatives of the American Gas Association (AGA), the American Society of Mechanical Engineers (ASME), and the National Fire Protection Association (NFPA). The *Code* was approved by the American National Standards Institute (ANSI) and is revised and updated as needed. The *Code* is listed as ANSI Z223.1 and NFPA 54. Copies may be obtained from ANSI and NFPA.

The *Uniform Plumbing Code* contains a section on fuel-gas piping.

The Southern Building Code Congress International has developed a *Standard Gas Code.*

If a question should arise regarding the installation or servicing of piping or appliances, the *local authority* (in certain codes the *administrative authority*) shall have final jurisdiction. Also, the word *shall* is a mandatory term.

Although all gas codes and regulations are designed to promote safety in the installation of piping and the servicing of gas-fueled equipment and appliances, and are, therefore, basically similar, the reader is advised to become thoroughly familiar with the gas code enacted in his or her area of work.

Capacity of Pipe of Different Diameters and Lengths in Cubic Feet per Hour with Pressure Drop of 0.3′ and Specific Gravity of 0.60

Length of Pipe, ft	*Iron-Pipe Sizes (IPS), in.*				
	½	¾	1	1¼	1½
15	76	172	345	750	1220
30	52	120	241	535	850
45	43	99	199	435	700
60	38	86	173	380	610
75		77	155	345	545
90		70	141	310	490
105		65	131	285	450
120			120	270	420
150			109	242	380
180			100	225	350
210			92	205	320
240				190	300
270				178	285
300				170	270
450				140	226
600				119	192

Length of Thread Required for Various Sizes of Gas Pipe

Size of Pipe, in.	*Approximate Length of Threaded Portion, in.*	*Approximate Number of Threads to Be Cut*
3/8	5/16	10
1/2	3/4	10
3/4	3/4	10
1	7/8	10
1 1/4	1	11
1 1/2	1	11
2	1	11
2 1/2	1 1/2	12
3	1 1/2	12
4	1 3/4	13

Sizing Gas Piping

The gas supply pipe should be sized according to volume of gas used and allowable pressure drop. The pipe must also be of adequate size to prevent undue pressure drop. The diameter of the supply pipe must at least equal that of the manual shutoff valve in the supply riser.

Formula for Sizing Gas Piping

To obtain the size of piping required for a certain unit, first determine the number of cubic feet of gas per hour consumed by the unit.

$$\text{Ft}^3 \text{ of gas/h} = \frac{\text{total Btu/h required by unit}}{\text{Btu/ft}^3 \text{ of gas}}$$

Example

Determine the size of the piping in a gas-fired boiler installation when the burner input is 155,000 Btu/h and the heating value per cubic foot of gas is 1000 Btu. Assume the distance from the gas meter to the boiler to be 75 feet.

By a substitution of values in the foregoing formula we obtain:

$$\text{Ft}^3 \text{ of gas/h} = {}^{155{,}000}/_{1000} = 155$$

A 1-inch pipe 75 feet long will handle 155 ft³/h with 0.3-inch pressure drop. The determination of gas pipe sizes for other appliance units can be made in a similar manner.

Example

Determine the pipe size of each section and outlet of a piping system with a designated pressure drop of 0.3 in the water column. The gas to be used has a specific gravity of 0.60 and a heating value of 1000 Btu per cubic foot.

Reference to the piping layout and a substitution of values for maximum gas demand for the various outlets (see the following figure) shows that the demand for individual outlets will be as follows:

$$\text{Outlet A} = {}^{55{,}000}/_{1000} = 55 \text{ ft}^3/\text{h}$$

$$\text{Outlet B} = {}^{3{,}000}/_{1000} = 3 \text{ ft}^3/\text{h}$$

$$\text{Outlet C} = {}^{65{,}000}/_{1000} = 65 \text{ ft}^3/\text{h}$$

$$\text{Outlet D} = {}^{116{,}000}/_{1000} = 116 \text{ ft}^3/\text{h}$$

From the piping layout in the following figure, it will be noted that the length of gas pipe from the meter to the most remote outlet (A) is 105 feet. This is the only distance measurement used. Reference to the figure indicates:

Outlet A, supplying 55 ft³/h requires ½-in. pipe.
Outlet B, supplying 3 ft³/h requires ½-in. pipe.
Outlet C, supplying 65 ft³/h requires ¾-in. pipe.
Outlet D, supplying 116 ft³/h requires ¾-in. pipe.

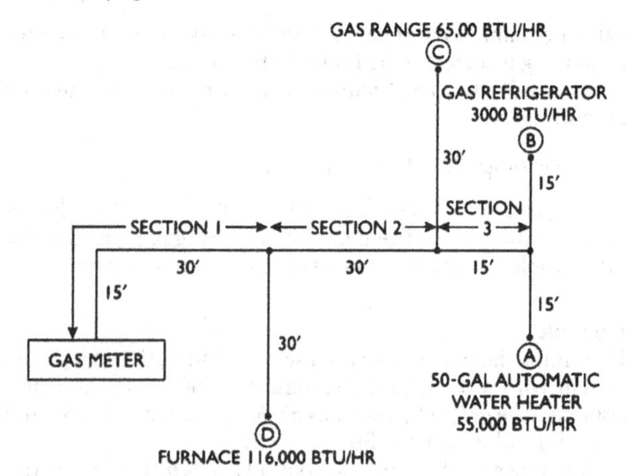

Diagram of a gas-piping layout.

Section 3, supplying outlets A and B, or 58 ft³/h, requires ½-in. pipe.

Section 2, supplying outlets A, B, and C, or 123 ft³/h, requires ¾-in. pipe.

Section 1, supplying outlets A, B, C, and D, or 239 ft³/h, requires 1-in. pipe.

The determination of sizes of gas pipe for other piping layouts may be made in a similar manner.

Installing Gas Piping

The installation and replacement of gas piping should be done only by qualified workers. Special attention should be taken to prevent leakage because of the volatile and highly flammable nature of the fuel.

The following recommendations are offered as a *guide* for installing or replacing gas piping. Think of it as a checklist. The furnace or boiler manufacturer's detailed installation instructions should always take precedence.

- All work should be done in accordance with the requirements of local codes and regulations governing the installation of gas piping.
- Locate the gas meter as close as possible to the point at which the gas service (supply) line enters the structure.
- Install a single gas line between the main gas supply and the furnace or boiler.
- Shut off the gas supply to the structure before beginning any work. Also, make sure the electricity is shut off at the main circuit breaker or fuse box.
- When installing a new system, size the pipes according to the volume of gas used and the allowable pressure drop.
- Use a piping material recommended by the local code. Never bend gas piping. Bending the pipe may cause the pipe walls to crack and leak gas.
- Use fittings for making turns in gas piping. Use 45° offsets to reduce friction to the flow of gas. Never use a 90° offset.
- Take all branch connections from the top or side of the horizontal pipes, never from the bottom.
- Make certain all pipes are adequately supported so that no unnecessary stress is placed on them.
- Use a suitable pipe sealant for joining pipe. Use a pipe compound resistant to LP gas on all threaded pipe connections in an LP gas installation.
- Install the supply riser and drop leg adjacent to and on either side of the furnace or boiler. Never install them

in front of the furnace or boiler where they will block access to the appliance.

- Install a manual main shutoff valve or plug cock on the riser at least 4 feet above the floor level and in accordance with the local codes and regulations.
- Install a tee fitting at the bottom of the riser to catch any foreign matter in the pipe. The bottom of the tee fitting should be plugged or capped.
- Extend the drip leg and cap it.
- Install a ground union joint between the tee and the furnace controls.
- Test for pressure and leaks. Check for gas leaks by applying a soap-and-water solution to the suspected area. *Never* use matches, candles, or any other flame to locate the leak.

GAS BURNERS

Gas Rate

Seconds Revolution	Size of Test Meter Dial				Seconds Revolution	Size of Test Meter Dial			
	½ ft³	1 ft³	2 ft³	5 ft³		½ ft³	1 ft³	2 ft³	5 ft³
	Cubic Feet per Hour					Cubic Feet per Hour			
10	180	360	720	1800	50	36	72	144	360
11	164	327	655	1636	51	35	71	141	353
12	150	300	600	1500	52	35	69	138	346
13	138	277	555	1385	53	34	68	136	340
14	129	257	514	1286	54	33	67	133	333
15	120	240	480	1200	55	33	65	131	327
16	112	225	450	1125	56	32	64	129	321
17	106	212	424	1059	57	32	63	126	316
18	100	200	400	1000	58	31	62	124	310
19	95	189	379	947	59	30	61	122	305
20	90	180	360	900	60	30	60	120	300
21	86	171	343	857	62	29	58	116	290
22	82	164	327	818	64	29	56	112	281
23	78	157	313	783	66	29	54	109	273
24	75	150	300	750	68	28	53	106	265
25	72	144	288	720	70	26	51	103	257
26	69	138	277	692	72	25	50	100	250
27	67	133	267	667	74	24	48	97	243
28	64	129	257	643	76	24	47	95	237
29	62	124	248	621	78	23	46	92	231
30	60	120	240	600	80	22	45	90	225

(continued)

(continued)

Seconds per Revolution	Size of Test Meter Dial: ½ ft³	1 ft³	2 ft³	5 ft³	Seconds per Revolution	Size of Test Meter Dial: ½ ft³	1 ft³	2 ft³	5 ft³
	Cubic Feet per Hour					Cubic Feet per Hour			
31	58	116	232	581	82	22	44	88	220
32	56	113	225	562	84	21	43	86	214
33	55	109	218	545	86	21	42	84	209
34	53	106	212	529	88	20	41	82	205
35	51	103	206	514	90	20	40	80	200
36	50	100	200	500	94	19	38	76	192
37	49	97	105	486	98	18	37	74	184
38	47	95	189	474	100	18	36	72	180
39	46	92	185	462	104	17	35	69	173
40	45	90	180	450	108	17	33	67	167
41	44	88	176	440	112	16	32	64	161
42	43	86	172	430	116	15	31	62	155
43	42	84	167	420	120	15	30	60	150
44	41	82	164	410	130	14	28	55	138
45	40	80	160	400	140	13	26	51	129
46	39	78	157	391	150	12	24	48	120
47	38	77	153	383	160	11	22	45	113
48	37	75	150	375	170	11	21	42	106
49	37	73	147	367	180	10	20	40	100

Courtesy Dunham-Bush, Inc.

Gas Burner Orifice Capacity

	Approximate Orifice Capacity (Btu/h)*	
Drill Size	**Natural Gas at 3 ½" W.C.**	**Propane Gas at 11" W.C.**
60	4650	13,000
59	4900	13,750
58	5150	14,450
57	5400	15,150
56	6300	17,600
55	7900	22,000
54	8800	24,750
53	10,400	28,800
52	11,750	33,000
51	13,000	36,700
50	14,250	40,000
49	15,500	43,500
48	16,750	47,200
47	17,900	50,200
46	19,000	53,500
45	19,500	55,000
44	21,500	60,050
43	23,000	
42	25,250	
41	26,750	
40	27,800	
39	28,600	
38	29,900	
37	31,100	
36	33,000	
35	35,000	

*Values determined using the following data: propane at 1.52 specific gravity/2500 Btu/ft^3 and natural gas at 0.62 specific gravity/1000 Btu/ft^3.

Courtesy The Trane Company

DUAL FUEL/CONVERSION GAS BURNER SPECIFICATIONS

Carlin Model G3B Residential Power Gas Burner

Model	*G3B Power Gas Burner*
Input	60,000 to 180,000 Btu/h
Fuel	Natural gas or propane gas
Maximum supply pressure	14 inches W. C.
Minimum supply pressure	5 inches W. C.
Manifold pressure	3.5 inches W. C.
Power	120 v, 60 Hz, single phase
Current	Approximately 8 A
Motor	1/50 hp, 3250
Gas valve power	24 v, 60 Hz
Ignition	Norton hot surface ignitor, 120 VAC
Control	Fenwal Model 05-32 Primary Control 7-second pre-purge ignitor warm-up Recycle on flame failure Factory-installed and wired-in control panel Panel includes valve-on light, transformer and relay

Carlin Residential Gas-Fired Burner (Model G3B Power Gas Burner) *(continued)*

Input MBH*	***Drill size Diameter (in.)***	
	Natural	Propane
60	#28	#33
70	#24	⅛
80	#20	#30
90	#17	#28
100	#14	#25
110	#11	#22
120	#7	#20
140	#3	#16
160	A	#12
180	D	#7

* *High altitude applications: Maximum input at sea level is 180 Btu/h. Reduce this capacity 4% per 1000 feet above sea level.*

Carlin Residential Model EZ-Gas Power Gas Burner

Model	*EZ-Gas Power Gas Burner*
Input	50,000 to 275,000 Btu/h
Fuel	Natural gas or propane gas
Maximum supply pressure	14 in. W.C.
Minimum supply pressure	5 in. W.C.
Manifold pressure	3.5 in. W.C.
Power	120 V, 60 Hz, single phase
Current	Approximately 2 A
Limit circuit input	120 V, 60 Hz
Motor	1/15 hp, 3450 rpm, 60 Hz
Motor frame	48 frame, "M" flange
Fuel valve power	120 VAC, 60 Hz
Ignition	Carlin Model 41800 Solid State Electronic Ignition; ignition voltage, 9000 V
Control	Carlin 60200FR Microprocessor Control 4-second trial for ignition (TFI) 1.3-second FFRT Prepurge and postpurge Serviceman reset protection Latch-up after three consecutive lockouts Interrupted duty ignition Recycle on flame failure

Carlin Residential Gas-Fired Burner (Model EZ-Gas Power Gas Burner) *(continued)*

Input	***Orifice Drill Size***		***Air Band***	
Btu/h	Natural Gas	Propane Gas	Slots	Approximate Setting*
50,000	#19 (0.166)	#32 (0.116)	1	0%
75,000	#11 (0.191)	#26 (0.147)	1	5%
100,000	#1 (0.228)	#16 (0.177)	1	20%
126,000	D (0.246)	#9 (0.196)	1	30%
150,000	J (0.277)	7⁄32 (0.219)	2	20%
175,000	N (0.302)	C (0.242)	2	30%
200,000	21⁄64 (0.328)	17⁄64 (0.266)	2	35%
225,000	T (0.358)	9⁄32 (0.281)	2	45%
250,000	X (0.397)	5⁄16 (0.312)	2	55%
275,000	7⁄16 (0.437)	R (0.339)	2	75%

* *Air band setting is approximate only. Follow burner manual and appliance manual instructions to adjust burner using combustion test instruments.*

Midco Economite RE4400DS and RE4400DSA Gas Burners

Model		*RE4400DS and RE4400DSA*
Firing rate*	Maximum MBH	400 MBH
	Minimum MBH	132 MBH
Required gas supply pressure	Natural gas	7.0 to 14.0 in. W.C.
	Propane	5.0 to 14.0 in. W.C.
Power		120 V, 60 Hz
Electronic control voltage		24 VAC
Ignition		Direct spark ignition
Tube diameter		4 in.
Tube length		9 in.
Recommended combustion chamber size (at maximum Btu/h)	Width	10 in.
	Length	16.5 in.
Air delivery (approximate air delivery at zero draft)		125 SCFM (standard cubic feet per minute)

**based on 1000 Btu/ft^3. natural gas and 2500 Btu/ft^3. propane. Derate burner for altitude over 2000 feet by 4% for each 1000 feet above sea level.*

Midco Economite RE4850A and RE4400 Gas Burners

Model		*RE4850A*	*RE4400*
Firing rate*	Maximum MBH	850 MBH	400 MBH
	Minimum MBH	280 MBH	132 MBH
Required gas supply pressure	Natural gas	7.0 to 14.0 in. W.C.	
	Propane	5.0 to 14.0 in. W.C.	
Power		120 V, 60 Hz	
Electronic control voltage		24 VAC	
Ignition		Intermittent and interrupted spark-ignited pilot and 100% shutoff	
Tube diameter		4 in.	
Tube length		9 in.	
Recommended combustion chamber size (at maximum Btu/h)	Width	10 in.	
	Length	16.5 in.	
Air delivery (approximate air delivery at zero draft)		125 SCFM (standard cubic feet per minute)	

**based on 1000 Btu/ft³ natural gas and 2500 Btu/ft³ propane. Derate burner for altitude over 2000 feet by 4% for each 1000 feet above sea level.*

Midco Economite E20B and E20BP Gas Conversion Burners

Model		*E20B*	*E20BP*
Firing rate*	Maximum	225 MBH	225 MBH
	Minimum	50 MBH	50 MBH
Required gas supply pressure	Natural gas	5.0 to 14.0 in. W.C.	5.0 to 14.0 in. W.C.
	Propane	5.0 to 14.0 in. W.C.	5.0 to 14.0 in. W.C.
Power		120 V, 60 Hz	120 V, 60 Hz
Pilot safety		Thermoelectric, 100% shutoff	Thermoelectric, 100% shutoff
Tube diameter		4 in.	4 in.
Tube length		8 in.	8 in.
Minimum combustion chamber size (at 225 MBH)	Width	7 in.	7 in.
	Length	11 in.	11 in.
	Diameter	10 in.	10 in.

**based on 1000 Btu/ft^3 natural gas and 2500 Btu/ft^3 propane. Derate burner for altitudes over 2000 feet by 4% for each 1000 feet of additional elevation.*

Midco Economite E20B and E20BP Gas Conversion Burners *(continued)*

Minimum combustion chamber size (at 150 MBH)	Width	6 in.	6 in
	Length	10 in.	10 in.
	Diameter	9 in.	9 in.
Air delivery (approximate air delivery at zero draft)		47 SCFM (standard cubic feet per minute)	47 SCFM (standard cubic feet per minute)

Midco Economite Legacy Series E200 and E300 Gas Conversion Burners

Model		*E200*	*E300*
Firing rate*	Maximum	200 MBH	300 MBH
	Minimum	70 MBH	90 MBH
Required gas supply pressure	Natural gas	6.0 to 14.0' in. W.C.	6.0 to 14.0 in. W.C.
	Propane	6.0 to 14.0' in. W.C.	6.0 to 14.0 in. W.C.
Power		120 V, 50/60 Hz, single phase (standard)	120 V, 50/60 Hz, single phase (standard)
		230 V, 50/60 Hz (contact factory)	230 V, 50/60 Hz (contact factory)
Current		2.0 A	3.0 A
Flame safety		24-volt Electronic flame safety with 100% shutoff, 30-second prepurge	24-volt electronic flame safety with 100% shutoff, 30-second prepurge
Tube diameter		4 in.	4 in.

(continued)

(continued)

Model		**E200**	**E300**
Tube length		7.50 in.	7.50 in.
Minimum combustion chamber size (at 100 MBH)	Width	8 in.	8 in.
	Height	8 in.	8 in.
Minimum combustion chamber size (at 200 MBH)	Width	10 in.	10 in.
	Height	10 in.	10 in.

** based on 1000 Btu/ft^3 for natural gas and 2500 Btu/ft^3 for propane at sea level.*

Midco Economite Legacy Series E200 and E300 Gas Conversion Burners *(continued)*

Minimum combustion chamber size (at 150 MBH)	Width	14 in.	14 in.
	Height	14 in.	14 in.
Air delivery (approximate air delivery at zero draft)		40.0 SCFM (standard cubic feet per minute)	60.0 SCFM (standard cubic feet per minute)
Main automatic valve		3-function redundant	3-function redundant

* *Based on 1000 Btu/ft.^{3} for natural gas and 2500 Btu/ft^3 for propane at sea level. Derate burner." Derate burner for altitudes over 2000 feet by 4% for each 1000 feet above sea level.*

Midco Economite Model 400-33 and F400-33 Gas Conversion Burners

Model		*400-33*	*F400-33*
Firing rate*	Maximum	400 MBH	700 MBH
	Minimum	185 MBH	300 MBH
Required gas supply pressure	Natural gas	5.0 to 14.0 in. W.C.	5.0 to 14.0 in. W.C.
	Propane	11.0 to 14.0 in. W.C.	11.0 to 14.0 in. W.C.
Power		120 V, 60 Hz	120 V, 60 Hz
Current			
Flame safety		Electronic flame safety with spark-ignited pilot and 100% shutoff	Electronic flame safety with spark-ignited pilot and 100% shutoff
Tube diameter			
Tube length			
Recommended combustion chamber size (at maximum MBH)	Width	10 in.	15 in.
	Length	16 ½ in.	25 in.

(continued)

(continued)

Model	***400-33***	***F400-33***
Air Delivery (approximate air delivery at zero draft)	85 SCFM (standard cubic feet per minute)	146 SCFM (standard cubic feet per minute)

* *Based on 1000 Btu/ft.^{3} for natural gas and 2500 Btu/ft^3 for propane. Derate burner for altitudes over 2000 feet by 4% for each 1000 feet above sea level.*

Midco Economite Models RE4850BA, RE4700BA, RE4400B, and RE4400BA Gas Burners

Model		*RE4850BA*	*RE4700BA*	*RE4400B*	*RE4400BA*
Firing rate*	Maximum	850 MBH	700 MBH	400 MBH	400 MBH
	Minimum	280 MBH	230 MBH	132 MBH	132 MBH
Required gas supply pressure	Natural gas	7.0 to 14.0 in. W.C.			
	Propane	5.0 to 14.0 in. W.C.			
Power		120 V, 60 Hz (for 50 Hz applications each, model will be derated 20%)			
Ignition control module voltage		24 VAC			
Flame safety		Electronic flame safety with spark-ignited pilot and 100% shutoff			
Tube diameter		4 in.			
Tube length		9 in.			
Recommended combustion chamber size (at maximum MBH)	Width	15 in.		10 in.	
	Length	30 in.	25 in.	16.5 in.	

(continued)

(Continued)

Model	**RE4850BA**	**RE4700BA**	**RE4400B**	**RE4400BA**
Air delivery (approximate air delivery at zero draft)	125 SCFM (standard cubic feet per minute)			

Based on 1000 Btu/ft^3. for natural gas and 2500 Btu/ft^3. for propane. Derate burner for altitudes over 2000 feet by 4% for each 1000 feet above sea level.

Midco Economite DS24A Conversion Gas Burner

Model		*DS24A*
Firing rate*	Maximum	225 MBH
	Minimum	50 MBH
Required gas supply pressure	Natural gas	5.0 to 14.0 in. W.C.
	Propane	5.0 to 14.0 in. W.C.
Power		120 V, 60 Hz
Flame safety		Direct spark ignition of main flame, electronic safety
Tube diameter		4 in.
Tube length		8 in.
Recommended minimum combustion chamber size (at 225 MBH)	Width	7 in.
	Length or diameter	11 in. length or 10 in. diameter
Recommended minimum combustion chamber size (at 150 MBH)	Width	6 in.
	Length or diameter	10 in. length or 9 in. diameter
Air delivery (approximate air delivery at zero draft)		47 SCFM (standard cubic feet per minute)

**based on 1000 Btu/ft.^{3} for natural gas and 2500 Btu/ft.^{3} for propane. Derate for altitudes over 2000 feet by 4% for each 1000 feet above sea level.*

Midco Economite 400B-02 and 400B-02P Conversion Gas Burners

Model		*400B-02*	*400B-02P*[*]
Firing rate[†]	Maximum	400 MBH	400 MBH
	Minimum	200 MBH	200 MBH
Required gas supply pressure	Natural gas	5.0 to 14.0 in. W.C.	N/A
	Propane	N/A	11.0 to 14.0 in. W.C.
Power		120 V, 60 Hz	120 V, 60 Hz
Pilot safety		Thermoelectric, 100% shutoff	Thermoelectric, 100% shutoff
Tube diameter		4 in.	
Tube length		9 in.	
Recommended minimum combustion chamber size (at 400 MBH)	Width	10 in.	
	Length or diameter	16 ½ in. length or 15 in. diameter	

Recommended minimum combustion chamber size (at 300 MBH)	Width	9 in.
	Length or diameter	14 in. length or 13 in. diameter
Recommended minimum combustion chamber size (at 200 MBH)	Width	7 in.
	Length or diameter	11 in. length or 10 in. diameter
Air delivery (approximate air delivery at zero draft)		85 SCFM (standard cubic feet per minute)

Model 400B-02P (propane) is equipped with a 4-function redundant main automaric valve.

†*Derate burner for altitudes over 2000 feet by 4% for each 1000 feet above sea level based on 1000 Btu/ft³ for natural gas and 2500 Btu/ft³ for propane.*

Riello 40 Series Gas Burners

Model	*On-Off*	*On-Off*	*On-Off*	*On-Off*	*On-Off*	*On-Off*	*On-Off*	*On-Off*	*2-Stage*	*2-Stage*
Type	*G120*	*G120*	*G200*	*G200*	*G400*	*G400*	*G750*	*G750*	*G900*	*G900*
Fuel type	Natural	Propane	Natural	Propane	Natural	Propane	Natural	Propane	Natural	Propane
Firing rate Btu/h × 1000	38–120	38–120	70–200	70–200	170–400	170–400	250–750	250–750	250–490* 360–900†	250–490* 360–900†
Manifold gas pressure (W.C.) minimum-maximum	.52–2.05	.51–1.87	.7–1.84	.9–2.96	.9–1.8	1.04–3.04	1.4–2.1	1.0–3.0	.55–1.1* 2.3–3.3†	1.7–3.4* 2.6–6.0†
Burner supply pressure (W.C.) minimum-maximum	4–10	8–13	4–10	8–13	4–10	8–13	7–14	8–13	7–14	8–13

Motor ratings	120 V 60 Hz 2.2A 3250 RPM	120 V 60 Hz 2.2A 3250 RPM	120 V 60 Hz 2.2A 3250 RPM	120 V 60 Hz 2.2A 3250 RPM	120 V 60 Hz 2.2A 3250 RPM	120 V 60 Hz 2.2A 3250 RPM	120 V 60 Hz 4.3A 3250 RPM	120 V 60 Hz 4.3A 3250 RPM	120 V 60 Hz 4.3A 3250 RPM	120 V 60 Hz 4.3A 3250 RPM

*Low fire
†High fire

Riello R Series Gas Burners

RS/M	Model	70	100	130	190
Capacity (MBtu/h)[1]		512–3084	570–4405	607–5545	1781–8673
	Voltage[2,3]	230–460–575/3/60	230–460–575/3/60	230–460–575/3/60	230–460–575/3/60
Combustion air motor	hp	1½	2½	3	5½
	Amperage draw	4.8/2.8/2.3	6.7/3.9/3.2	8.8/5.1/4.1	15.8/9.1/7.3
Sound levels	dBA	75.0	77.0	78.5	83.1
Maximum chamber pressure (W.C.) @ high fire		2.0	2.8	1.6	0.8
Minimum gas pressure (W.C.)[4,5]		7.0	7.0	8.0	7.0

(1) Ratings based on 1000 Btu/ft^3 and 0.64 specific gravity.
(2) Other voltages available on request.
(3) A separate 120/1/60 control voltage supply is required.
(4) Based on natural gas.
(5) Honeywell, Dungs, and Siemens gas trains available.

RESIDENTIAL AND LIGHT-COMMERCIAL OIL BURNER SPECIFICATIONS

Beckett NX Residential Oil Burner

Model		*Beckett NX*
Firing rate	gph	0.40–1.75
	Btu/h input	56,000–280,000
Heating oil		No. 1 or No. 2 heating oil (ASTMD396) only
Power supply		120 v 60 Hz single phase
Motor operating load (rated)		2.0 A maximum
Motor rpm (rated)		3450
Fuel unit (pump)		Single stage
Ignition		Continuous duty, solid-state igniter
Dimensions (standard)	Height	12½ ft
	Width	15 ft
	Depth	9¼ ft

Beckett AF Residential Oil Burner

Model		*Beckett AF*
Firing rate	gph	0.40 to 3.00
	Btu/h input	56,000 to 420,000
Heating oil		No. 1 or No. 2 heating oil (ASTM D396) only
Power supply		120 V 60 Hz single phase
Motor operating load (rated)		5.8 A (maximum)
Motor rpm (rated)		3450 rpm, NEMA M-flange, manual-reset overload protection
Fuel unit (pump)		Single stage or two stage
Ignition		10,000 V/23 MA secondary, continuous-duty shielded transformer
Dimensions (standard)	Height	11½ in.
	Width	12⅞ in.
	Depth	6⁹⁄16 in.

Beckett AFG Residential Oil Burner

Model		*Beckett AFG*
Firing rate	gph	0.40 to 3.00
	Btu/h input	56,000 to 420,000
Heating oil		No. 1 or No. 2 heating oil (ASTM D396) only
Power supply		120 V, 60 Hz single phase
Motor operating load (rated)		5.8 A maximum
Motor rpm (rated)		3450 rpm, NEMA M-flange, manual-reset overload protection
Fuel unit (pump)		Single Stage or Two Stage
Ignition		10,000 V/23 mA secondary, continuous-duty shielded transformer
Dimensions (standard)	Height	11½ in.
	Width	12⅞ in.
	Depth	6⁹⁄₁₆ in.

Beckett AFII Residential Oil Burners

Model		*Beckett AFII 85*	*Beckett AFII 150*
Firing rate	gph	0.40 to 0.85	0.75 to 1.35
	Btu/h input	56,000 to 119,000	105,000 to 189,000
Heating oil		No. 1 or No. 2 heating oil (ASTM D396) only	No. 1 or No. 2 heating oil (ASTM D396) only
Power supply		120 V, 60 Hz single phase	120 V, 60 Hz single phase
Motor operating load (rated)		5.8 A	5.8 A
Motor rpm (rated)		3450 rpm, NEMA M-flange, manual-reset overload protection	3450 rpm, NEMA M-flange, manual-reset overload protection
Fuel unit (pump)		Single stage or two stage	Single stage or two stage
Ignition		10,000 V/23 mA secondary, continuous-duty shielded transformer	10,000 V/23 mA secondary, continuous-duty shielded transformer
Dimensions (standard)	Height	13¼ in.	13¼ in.
	Width	14½ in.	14½ in.
	Depth	6⅝ in.	6⅝ in.

Carlin Residential EZ-Pro Advanced Oil Burners

Model	*EZ-1*	*EZ-2*	*EZ-3*
Input	0.5–1.65 gph	1.50–2.25 gph	2.00–2.50 gph
Heating oil	No. 1 or No. 2 fuel oil; Canada: No. 1 fuel oil or No. 2 furnace oil		
Power supply	120 V, 60 Hz single phase		
Motor operating load (rated)	Approximately 5.8 A		
Motor (PSC)	⅙ hp, 3450 rpm		
Motor frame	48-frame, "M" flange		
Ignition	Carlin Model 41000 Solid Electronic Ignition; 14,000 Vol		
Control	Carlin Model 60200 Microprocessor Control 15-second trial for ignition (TFI); 1.3-second FFRT; prepurge and postpurge (10 seconds); serviceman reset protection (latch-up after three consecutive lockouts)		
Oil valve power	120 V, 60 Hz		
Nozzle line heater	120 V, 60 Hz		
Alarm contacts (isolated)	24 VAC/VDC, 2 A		
Operating temperature limits	Maximum ambient temperature 104°F (40°C)		

Carlin Model 102CRD, 99FRD, and 100CRD Residential Oil Burners

Model	*102CRD*	*99FRD*	*100CRD*
Input	2.00–3.50 gph (2⅝" air cone) 3.00–4.50 gph (3" Air Cone)	0.50–3.00 gph	0.50–2.25 gph
Heating oil	No. 1 or No. fuel oil; Canada: No. 1 fuel oil or No. 2 furnace oil		
Power supply	120 V, 60 Hz single phase		
Motor (PSC)	⅙ hp, 3450 rpm		
Motor operating load (rated)	Approximately 6.0 A		
Motor frame	48-frame, "M" flange		
Ignition	Carlin Model 41000 Solid State Electronic Ignition 14,000 V		
Optional oil valve power	120 V, 60 HZ		
Optional nozzle line heater	120 V, 60 HZ		
Limit circuit input	120 V, 60 HZ		

Riello R-35 Series Oil Burners

Model		***R35.3***	***R35.5***
	gph	0.60–1.10	0.75–1.60
Firing rate	Btu/h × 1000	84–154	105–224
	kW/h	24–45	30–65
Usable heating oil		No heavier than No. 2*	
Power supply		120 V, 60 Hz	
Motor load (rated)		2.2 A	
Motor rpm (rated)		3250	
Pump pressure range		120–200 psi	

**Optional pump required for kerosene.*

Riello 40 Series Oil Burners

Model		F3	F5	F10	F15	F20
	gph	0.50–0.95	0.75–1.65	1.45–2.95	2.55–5.75	3.50–6.40
Firing rate	Btu/h × 1000	70–133	105–231	203–413	357–805	490–896
	kW/h	20–39	30–67	59–121	104–236	143–263
Usable heating oil		No heavier than No. 2 (optional pump required for kerosene)				
Motor ratings		120 V, 60 Hz, 3250 rpm				
			2.2 A			4.3 A
Pump pressure range		120–200 psi				

Riello BF40 Balanced Flue Model Oil Burner

Model		*BF3*	*BF5*
Firing rates	gph	0.50–0.95	0.75–1.65
	Btuh × 1000	70–133	105–231
	kWh	20–39	30–67
Usable heating oil		No heavier than No. 2 (optional pump required for kerosene)	
Motor ratings		120 v 60 Hz 3250 rpm	
		1.6 A	1.7 A
Pump pressure range		120–200 psi	

BURNER NOZZLES

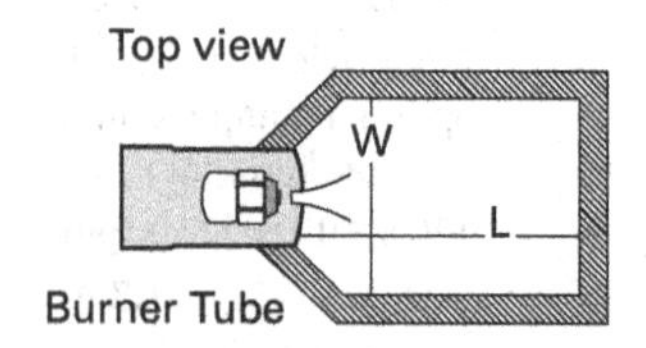

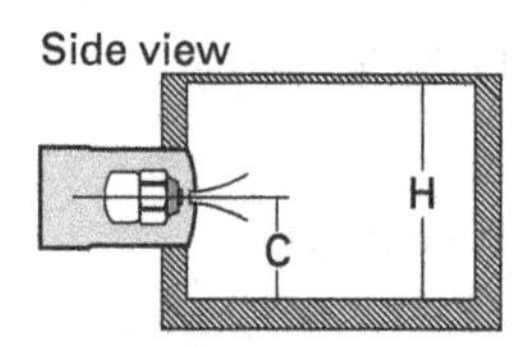

Recommended Combustion Chamber Dimensions

		Square or Rectangular Combustion Chamber				
Nozzle Size or Rating (gph)	***Spray Angle***	***L Length in.***	***W Width in.***	***H Height in.***	***C Nozzle Height in.***	***Round Chamber (diameter in in.)***
0.50–0.65	80°	8	8	11	4	9
0.75–0.85	60°	10	8	12	4	*
	80°	9	9	13	5	10
1.00–1.10	45°	14	7	12	4	*
	60°	11	9	13	5	*
	80°	10	10	14	6	11
1.25–1.35	45°	15	8	11	5	*
	60°	12	10	14	6	*
	80°	11	11	15	7	12
1.50–1.65	45°	16	10	12	6	*
	60°	13	11	14	7	*
	80°	12	12	15	7	13

(continued)

		Square or Rectangular Combustion Chamber				
Nozzle Size or Rating (gph)	**Spray Angle**	**L Length in.**	**W Width in.**	**H Height in.**	**C Nozzle Height in.**	**Round Chamber (diameter in in.)**
1.75–2.00	45°	18	11	14	6	*
	60°	15	12	15	7	*
	80°	14	13	16	8	15
2.25–2.50	45°	18	12	14	7	*
	60°	17	13	15	8	*
	80°	15	14	16	8	16
3.0	45°	20	13	15	7	*
	60°	19	14	17	8	*
	80°	18	16	18	9	17

**Recommend oblong chamber for narrow sprays.*

These dimensions are for average conversion burners. Burners with special firing heads may require special chambers.

Higher back wall, flame baffle or corbelled back wall increase efficiency on many jobs.

Combustion chamber floor should be insulated on conversion jobs.

For larger nozzle sizes use the same approximate proportions and 90 in. of floor area per 1 gph.

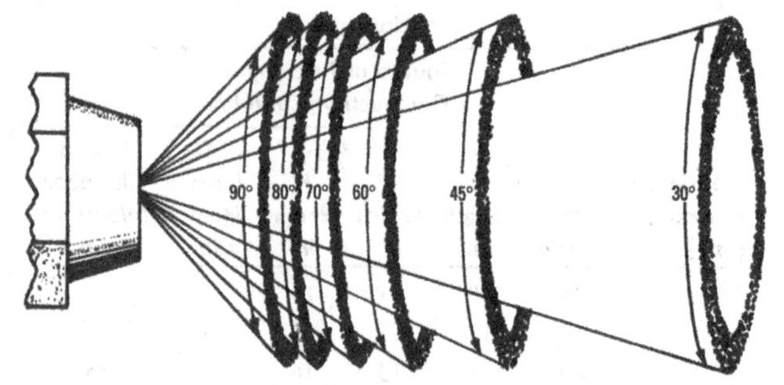

Various spray angles for oil burner nozzles.
(Courtesy Wayne Home Equipment Co., Inc.)

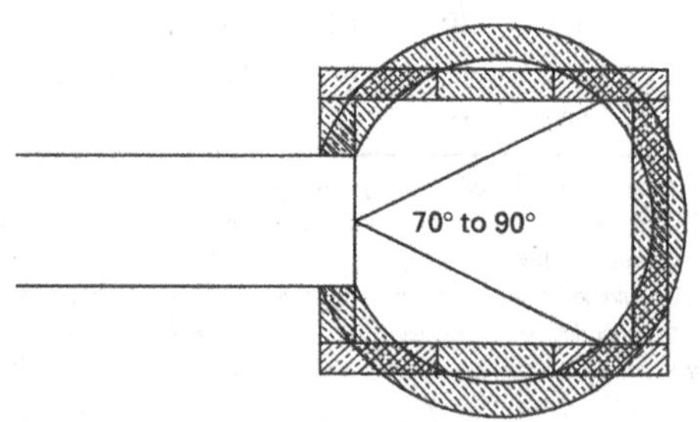

70° TO 90° SPRAY ANGLES FOR ROUND OR SQUARE CHAMBERS

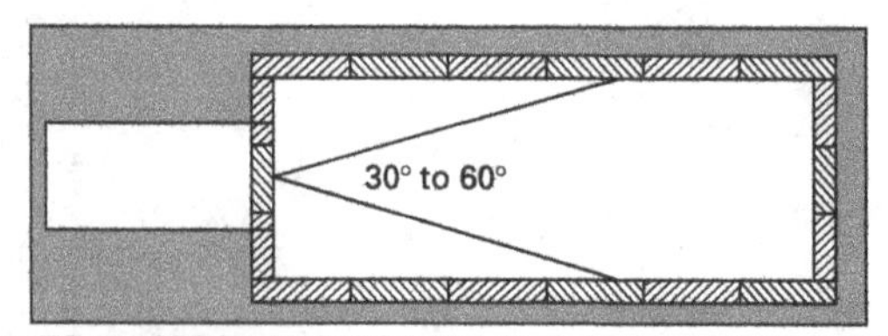

30° TO 60° SPRAY ANGLES FOR LONG. NARROW CHAMBERS

Recommended spray angles for round, square, and rectangular combustion chambers.

Proper Burner Nozzle Flow Rates

The proper size nozzle for a given burner unit is sometimes stamped on the nameplate of the unit. The following guidelines may be used for determining the proper flow rates:

For a unit rating given in Btu per hour input, the proper flow rate and nozzle size may be determined by using the following formula:

gph = Btu input / 140,000

For a unit rating given in Btu output, use the following formula:

gph = Btu output / (efficiency %) × 140,000

For a steam heating system in which the total square feet of steam radiation (including piping) is known, use the following formula to determine proper flow rate and nozzle size:

gph = total square feet of steam × 240 / (efficiency %) × 140,000

For a hot water (hydronic) heating system operating at 180° in which the total square feet of radiation (including piping) is known, use the following formula:

gph = total square feet of hot water × 165 / (efficiency %) × 140,000

Effects of Pressure on Nozzle Flow Rate

Nozzle Rating at 100 psi	*Nozzle, Flow Rates in gph (approximate)*					
	120 psi	*145 psi*	*160 psi*	*175 psi*	*200 psi*	*300 psi*
0.40	0.44	0.48	0.51	0.53	0.57	0.69
0.50	0.55	0.60	0.63	0.66	0.71	0.87
0.60	0.66	0.72	0.76	0.79	0.85	1.04
0.65	0.71	0.78	0.82	0.86	0.92	1.13
0.75	0.82	0.90	0.95	0.99	1.06	1.30
0.85	0.93	1.02	1.08	1.12	1.20	1.47
0.90	0.99	1.08	1.14	1.19	1.27	1.56
1.00	1.10	1.20	1.26	1.32	1.41	1.73
1.10	1.20	1.32	1.39	1.46	1.56	1.91
1.20	1.31	1.44	1.52	1.59	1.70	2.08
1.25	1.37	1.51	1.58	1.65	1.77	2.17
1.35	1.48	1.63	1.71	1.79	1.91	2.34
1.50	1.64	1.81	1.90	1.98	2.12	2.60
1.65	1.81	1.99	2.09	2.18	2.33	2.86
1.75	1.92	2.11	2.21	2.32	2.47	3.03
2.00	2.19	2.41	2.53	2.65	2.83	3.46
2.25	2.46	2.71	2.85	2.98	3.18	3.90
2.50	2.74	3.01	3.16	3.31	3.54	4.33
2.75	3.01	3.31	3.48	3.64	3.89	4.76
3.00	3.29	3.61	3.79	3.97	4.24	5.20
3.25	3.56	3.91	4.11	4.30	4.60	5.63
3.50	3.83	4.21	4.43	4.63	4.95	6.06
4.00	4.38	4.82	5.06	5.29	5.66	6.93
4.50	4.93	5.42	5.69	5.95	6.36	7.79
5.00	5.48	6.02	6.32	6.61	7.07	8.66

(continued)

(continued)

Nozzle Rating at 100 psi	***Nozzle, Flow Rates in gph (approximate)***					
	120 psi	***145 psi***	***160 psi***	***175 psi***	***200 psi***	***300 psi***
5.50	6.02	6.62	6.96	7.28	7.78	9.53
6.00	6.57	7.22	7.59	7.94	8.49	10.39
6.50	7.12	7.83	8.22	8.60	9.19	11.26
7.00	7.67	8.43	8.85	9.26	9.90	12.12
7.50	8.22	9.03	9.49	9.92	10.61	12.99
8.00	8.76	9.63	10.12	10.58	11.31	13.86
8.50	9.31	10.24	10.75	11.24	12.02	14.72
9.00	9.86	10.84	11.38	11.91	12.73	15.59
9.50	10.41	11.44	12.02	12.57	13.44	16.45
10.00	10.95	12.04	12.65	13.23	14.14	17.32
11.00	12.05	13.25	13.91	14.55	15.56	19.05
12.00	13.15	14.45	15.18	15.87	16.97	20.78
13.00	14.24	15.65	16.44	17.20	18.38	22.52
14.00	15.34	16.86	17.71	18.52	19.80	24.25
15.00	16.43	18.06	18.97	19.84	21.21	25.98
16.00	17.53	19.27	20.24	21.17	22.63	27.71
18.00	19.72	21.67	22.77	23.81	25.46	31.18
20.00	21.91	24.08	25.30	26.46	28.28	34.64
22.00	24.10	26.49	27.83	29.10	31.11	38.11
24.00	26.29	28.90	30.36	31.75	33.94	41.57
26.00	28.48	31.31	32.89	34.39	36.77	45.03
28.00	30.67	33.72	35.42	37.04	39.60	48.50
30.00	32.86	36.12	37.95	39.69	42.43	51.96
32.00	35.05	38.53	40.48	42.33	45.25	55.43
35.00	38.34	42.15	44.27	46.30	49.50	60.62

(continued)

Nozzle Rating at 100 psi	Nozzle, Flow Rates in gph (approximate) 120 psi	145 psi	160 psi	175 psi	200 psi	300 psi
40.00	43.82	48.17	50.60	52.92	56.57	69.28
45.00	49.30	54.19	56.92	59.53	63.64	77.94
50.00	54.77	60.21	63.25	66.14	70.71	86.60

Courtesy Delavan Spray Technologies

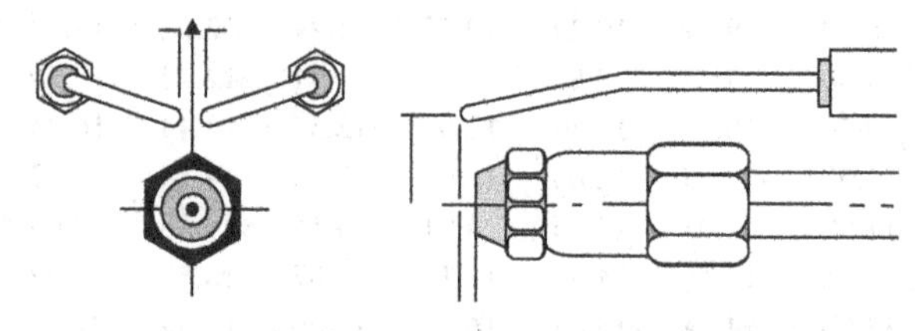

Recommended Electrode Settings

Nozzle	GPH	A	B	C
45°	0.75 to 4.00	⅛" to 3⁄16"	½" to 9⁄16"	¼"
60°	0.75 to 4.00	⅛" to 3⁄16"	9⁄16" to ⅝"	¼"
70°	0.75 to 4.00	⅛" to 3⁄16"	9⁄16" to ⅝"	⅛"
80°	0.75 to 4.00	⅛" to 3⁄16"	9⁄16" to ⅝"	⅛"
90°	0.75 to 4.00	⅛" to 3⁄16"	9⁄16" to ⅝"	0

OIL STORAGE TANKS

The location of the oil supply tank is subject to local regulations. These must be consulted before a *new* tank is installed.

The supply tank can be located inside or outside the building, above the level of the oil burner or below it. Furthermore, outside tanks can be located underground or above it. It is recommended that larger supply tanks be located outside and underground.

The following figures illustrate four ways to locate oil supply tanks. When installing a fuel tank, the following suggestions should prove helpful:

1. The filler pipe should be a *minimum* of 2 inches in diameter; the vent pipe, 1¼ inches.
2. Use wrought-iron pipe with malleable-iron fittings for both the filler and vent pipes.
3. Coat only the *male* thread of the pipe with a pipe compound suitable for use with oil burning equipment.
4. Oil supply lines between the oil supply tank and the oil burner should be made of copper tubing (diameters will vary depending on local regulations, pipe length, and the specifications of the oil burner being used).
5. Use a floor-level tee if the oil supply lines run overhead.
6. The oil burners for the water heater and furnace (or boiler) may be connected to a common feed line in conventional gravity feed installations.
7. No return line is necessary when the supply tank is installed above the level of the oil burner and the oil is fed by gravity to the burner.
8. Use a single-line system if the oil is gravity fed from the supply tank to the burner.
9. Use a two-line system if the oil tank is buried and below the level of the burner.

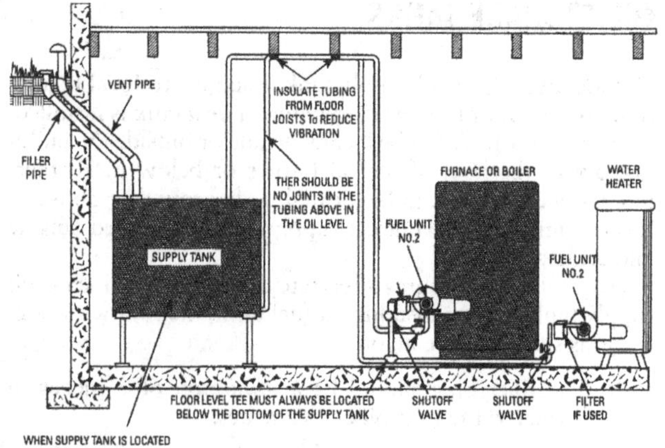

Inside tank installation.

(Courtesy Sundstrand Hydraulics)

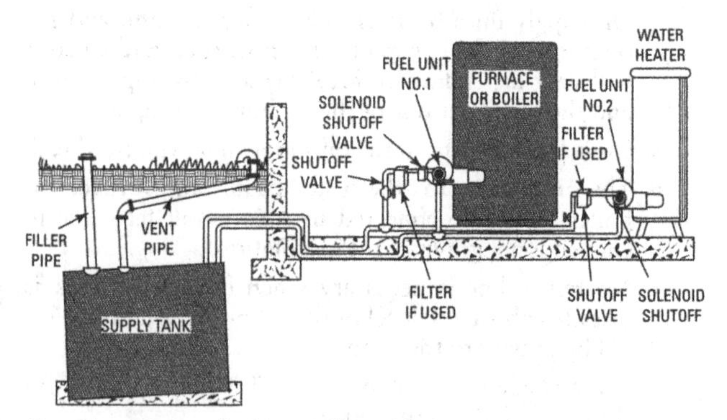

Outside tank installation involving lift.

(Courtesy Sundstrand Hydraulics)

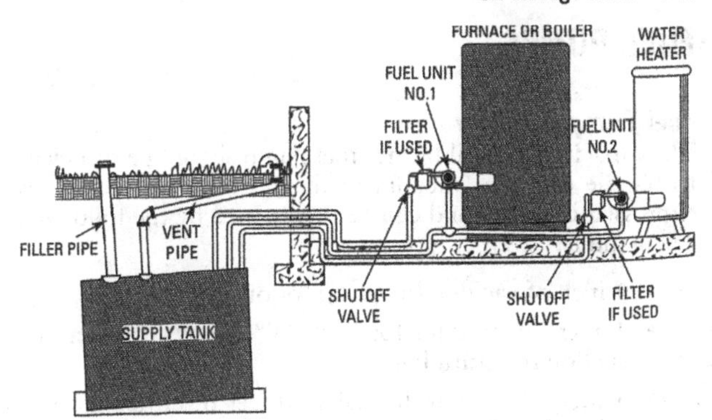

Outside tank installation with the tank located below the level of the oil burner.

(Courtesy Sundstrand Hydraulics)

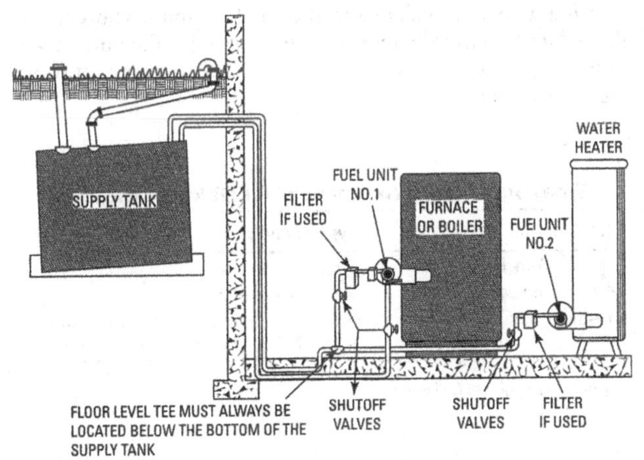

Outside tank installation with the tank located above the level of the oil burner.

(Courtesy Sundstrand Hydraulics)

FUEL PUMPS

Fuel Pump Capacity

The capacity of an oil burner fuel pump should be sufficient to handle the total vacuum in the system. The vacuum is expressed in inches and can be determined by the following calculations:

- 1 inch of vacuum for each foot of lift
- 1 inch of vacuum for each 90° elbow in either the suction or return lines
- 1 inch of vacuum for each 10 feet of horizontal run (3/8-inch OD line)
- 1 inch of vacuum for each 20 feet of horizontal run (1/2-inch OD line)

After you have calculated the total vacuum, you can use these data to select the most suitable pump for the burner. The following table lists various vacuums and suggests appropriate pump capacities.

Types of Pumps Recommended for Different Vacuums

Total Vacuum	*Type of Pump*
Up to a 3-inch vacuum	Single-stage pump
4–13-inch vacuum	Two-stage pump
14-inch vacuum or more	Single-stage pump for the burner and a separate lift pump with a reservoir

(Courtesy National Fuel Oil Institute)

Adjusting Fuel Pump Pressure

The oil pressure regulator on the fuel pump is generally factory-set to give nozzle oil pressures of 100 psig. The firing rate is indicated on the pump nameplate. The firing rate can be obtained with standard nozzles by inserting a gauge in the pump gauge port and turning the adjusting screw clockwise to increase pressure or counterclockwise to decrease it. Do not exceed the maximum recommended pressure (psig).

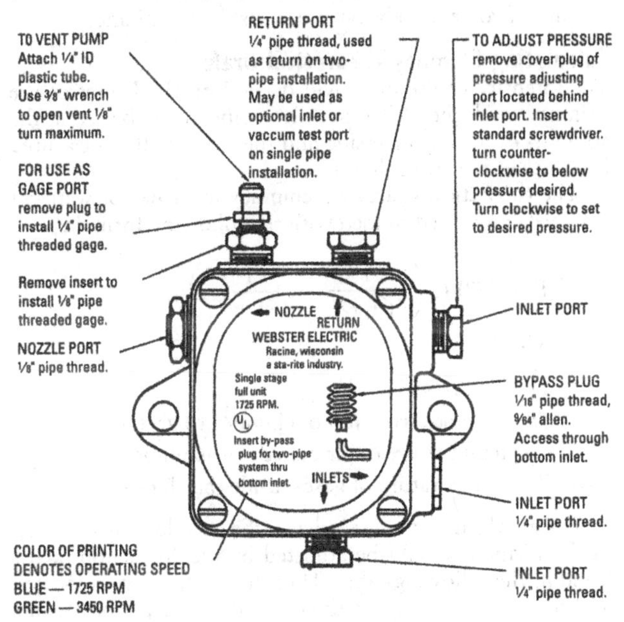

Webster model M series fuel pump.

(Courtesy Webster Electrical Co., Inc.)

CHIMNEYS AND FIREPLACES

A properly designed and constructed chimney is essential to a fireplace because it provides the draft necessary to remove the smoke and flue gases. The motive power that produces this natural draft is the slight difference in weight between the column of rising hot flue gases *inside* the chimney and the column of colder and heavier air *outside* the chimney.

Calculating Chimney Theoretical Draft

The intensity of the draft will depend on the height of the chimney and the difference in temperature between the columns of air on the inside and the outside. It is measured in inches of a water column.

The theoretical draft of a chimney in inches of water at sea level can be determined with the following formula:

$$D = 7.00\,H\left(\frac{1}{461 + T} - \frac{1}{461 + T_1}\right)$$

Where

D = theoretical draft
H = distance from top of chimney to grates
T = temperature of air outside the chimney
T_1 = temperature of gases inside the chimney

For an altitude *above* sea level, the calculations obtained in the formula should be adjusted by the correction factor listed in the following table. Thus, if the structure is located at an altitude of approximately 1000 feet, the results obtained in the formula should be multiplied by the correction factor 0.966 to obtain the correct draft for that altitude.

Chimney Draft Correction Factors for Altitudes above Sea Level

Altitude (in feet)	*Correction Factor*
1000	0.966
2000	0.932
3000	0.900
5000	0.840
10,000	0.694

Chimney Details

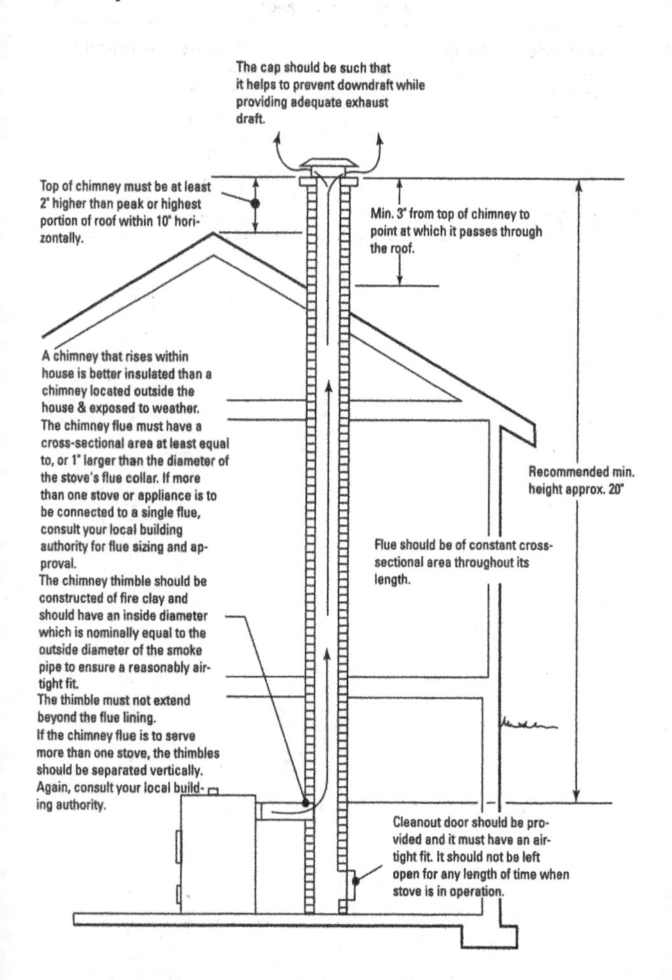

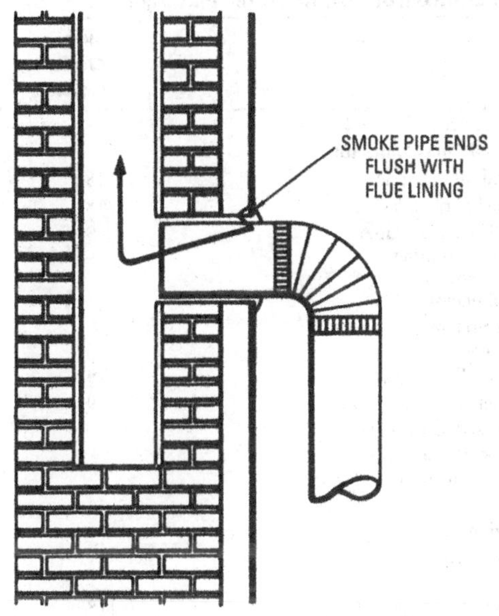

Connection of smoke pipe to chimney flue.

Chimney Connector and Vent Connector Clearance from Combustible Materials

Description of Appliance	*Minimum Clearance, Inches**
Residential Appliances	
Single-Wall, Metal Pipe Connector	
Electric, gas, and oil incinerators	18
Oil and solid-fuel appliances	18
Oil appliances listed as suitable for use with Type-L venting system, but only when connected to chimneys	9
Type-L Venting-System Piping Connectors	
Electric, gas, and oil incinerators	9
Oil and solid-fuel appliances	9
Oil appliances listed as suitable for use with Type-L venting systems	[a]
Commercial and Industrial Appliances	
Low-Heat Appliances	
Single-Wall, Metal Pipe Connectors	
Gas, oil, and solid-fuel boilers, furnaces, and water heaters	18
Ranges, restaurant type	18
Oil unit heaters	18
Other low-heat industrial appliances	18
Medium-Heat Appliances	
Single-Wall, Metal Pipe Connectors	
All gas, oil, and solid-fuel appliances	36

**These clearances apply except if the listing of an appliance specifies different clearances, in which case the listed clearance takes precedence.*

[a]If listed Type-L venting-system piping is used, the clearance may be in accordance with the venting-system listing.

If listed Type-B or Type-L venting-system piping is used, the clearance may be in accordance with the venting-system listing.

The clearances from connectors to combustible materials may be reduced if the combustible material is protected in accordance with Table 1C.

Courtesy National Oil Fuel Institute

Fireplace Details

Recommended Dimensions for a Finished Masonry Fireplace*

Opening							
Width, w	Height, h	Depth, d	Minimum Back (Horizontal), c	Vertical Back Wall, a	Inclined Back Wall, b	Outside Dimensions of Standard Rectangular Flue Lining	Inside Diameter of Standard Round Flue Lining
(in)	*(in)*	*(in)*	*(in)*	*(in)*	*(in)*	*(in)*	*(in)*
24	24	16–18	14	14	16	8½ by 8½	10
28	24	16–18	14	14	16	8½ by 8½	10
24	28	16–18	14	14	20	8½ by 8½	10
30	28	16–18	16	14	20	8½ by 13	10
36	28	16–18	22	14	20	8½ by 13	12
42	28	16–18	28	14	20	8½ by 18	12
36	32	18–20	20	14	24	8½ by 18	12
42	32	18–20	26	14	24	13 by 13	12
48	32	18–20	32	14	24	13 by 13	15
42	36	18–20	26	14	28	13 by 13	15
48	36	18–20	32	14	28	13 by 18	15
54	36	18–20	38	14	28	13 by 18	15
60	36	18–20	44	14	28	13 by 18	15
42	40	20–22	24	17	29	13 by 13	15
48	40	20–22	30	17	29	13 by 18	15
54	40	20–22	36	17	29	13 by 18	15
60	40	20–22	42	17	29	18 by 18	18
66	40	20–22	48	17	29	18 by 18	18
72	40	22–28	51	17	29	18 by 18	18

Letters at Heads of Columns Refer to the figures that follow.

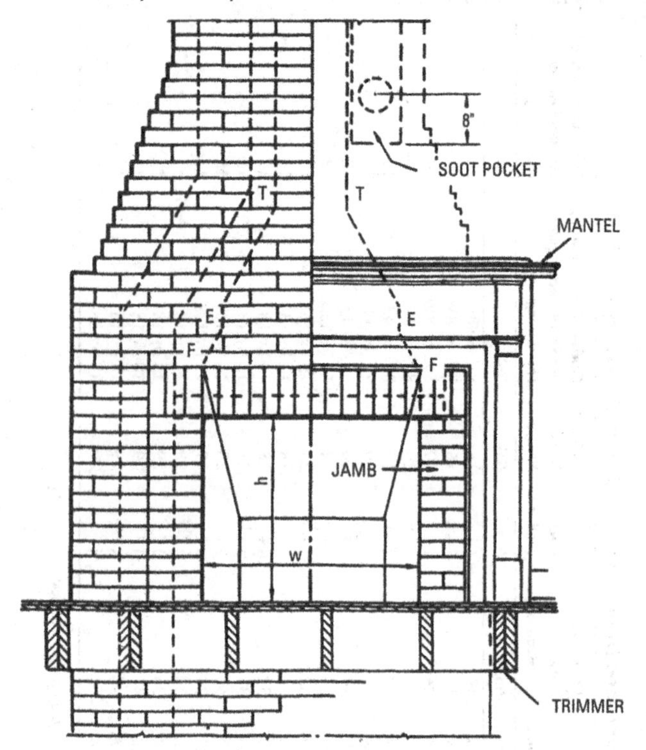

Fireplace elevation.

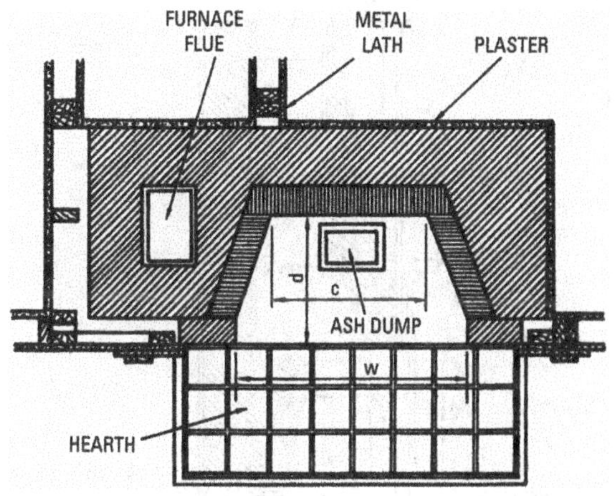

Fireplace plan.

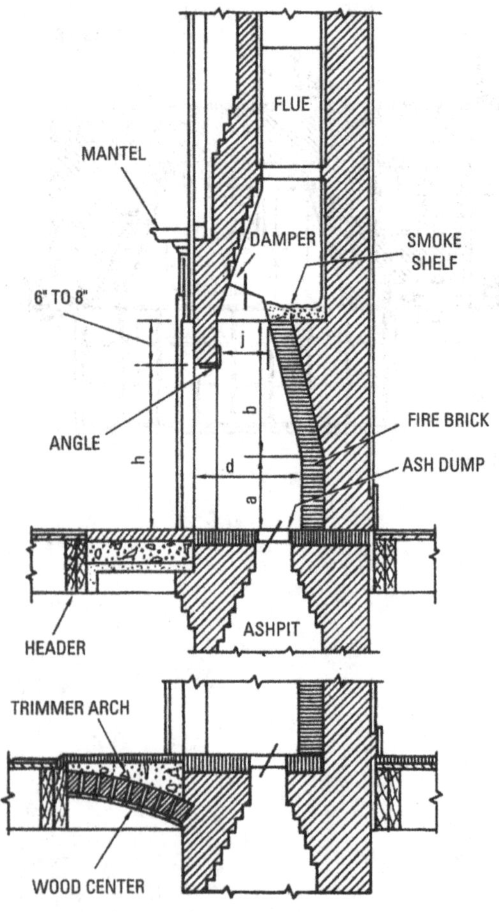

Fireplace sectional views illustrating two types of hearth construction.

STEAM HEATING SYSTEMS

Some Facts about Steam

- A cubic inch of water evaporated under ordinary atmospheric pressure (14.7 psig) will be converted into approximately one cubic foot of steam. This cubic foot of steam exerts a mechanical force equal to that needed to lift 1955 pounds 1 foot.
- The specific gravity of steam at atmospheric pressure is 0.462 that of air at 32°F and 0.0006 that of water at the same temperature. Therefore, 28.21 cubic feet of steam at atmospheric pressure weighs 1 pound, and 12.387 cubic feet of air weighs one pound.
- Each nominal horsepower of boilers requires from 4 to 6 gallons of water per hour.
- Good boilers will evaporate from 10 to 12 pounds of water per pound of coal.
- One square foot of grate surface will consume from 10 to 12 pounds of hard coal or from 18 to 20 pounds of soft coal per hour with natural draft. With forced drafts these amounts can be doubled.
- In calculating the horsepower of boilers, allow about 11 ½ square feet of heating surface per horsepower.
- The standard unit of the horsepower of boilers (1915 Power Test Code, ASME) is as follows: one boiler horsepower is equivalent to the evaporation, from 212°F feed water, of 34 ½ pounds of water into dry saturated steam at 212°F.
- Steam at a given temperature is said to be saturated when it is of maximum density for that temperature. Steam in contact with water is saturated steam.
- Steam that has water in the form of small drops suspended in it is called *wet*, or supersaturated, steam. If

wet steam is heated until all the water suspended in it is evaporated, it is said to be *dry* steam.

- If dry saturated steam is heated when not in contact with water, its temperature is raised and its density is diminished or its pressure is raised. The steam is then said to be *superheated*.

Sizing Steam Boilers

Steam boilers are sized by first calculating the square feet of steam produced by each radiator in the structure and then adding all the radiator calculations together. Then, refer to the sizing literature for the proposed boiler make and model. In the "square feet of steam column" for the boiler, find the figure that comes closest to the total square feet of steam for the structure (that is, the total of the calculations for each of the radiators in the structure). If the closest figure in the column is less than the total calculation of feet of steam for the structure, always choose the larger size. Never choose a boiler that delivers a lower output (square feet of steam) than the total system requirement.

Size of Steam Mains

Radiation, (ft^2)	One-Pipe Work, (in)	Two-Pipe Work, (in)
125	1½	1¼ × 1
250	2½	1½ × 1¼
400	3	2 × 1½
650	3½	2½ × 2
900	4	3 × 2½
1250	4½	3½ × 3
1600	5	4 × 3½
2050	5½	4½ × 4
2500	6	5 × 4½
3600	7	6 × 5
5000	8	7 × 6
6500	9	8 × 6
8100	10	9 × 6

Sizing Traps for Steam Mains—Condensation Load in Pounds per Hour per 1000 Feet of Insulated Steam Main*—Ambient Temperature 70°F—Insulation 80% Efficient

Steam Pressure (psig)	Main Size 2"	2½"	3"	4"	5"	6"	8"	10"	12"	14"	16"	18"	20"	24"	0°F† Correction Factor
10	6	7	9	11	13	16	20	24	29	32	36	39	44	53	1.58
30	8	9	11	14	17	20	26	32	38	42	48	51	57	68	1.50
60	10	12	14	18	24	27	33	41	49	54	62	67	74	89	1.45
100	12	15	18	22	28	33	41	51	61	67	77	83	93	111	1.41
125	13	16	20	24	30	36	45	56	66	73	84	90	101	121	1.39
175	16	19	23	26	33	38	53	66	78	86	98	107	119	142	1.38
250	18	22	27	34	42	50	62	77	92	101	116	126	140	168	1.36
300	20	25	30	37	46	54	68	85	101	111	126	138	154	184	1.35
400	23	28	34	43	53	63	80	99	118	130	148	162	180	216	1.33
500	27	33	39	49	61	73	91	114	135	148	170	185	206	246	1.32
600	30	37	44	55	68	82	103	128	152	167	191	208	232	277	1.31

*Chart loads represent losses due to radiation and convection for standard steam.

† For outdoor temperature of 0°F, multiply load value in table for each main size by correction factor corresponding to steam pressure.

(Courtesy Spirax Sarco Co.)

Effect of Back Pressure on Steam Trap Capacity (percentage reduction in capacity)

% Back Pressure	Inlet Pressure psig 5	25	100	200
25	6	3	0	0
50	20	12	6	5
75	38	30	25	23

(Courtesy Spirax Sarco Co.)

Removing Air from Steam Lines

Steam cannot be maintained at its saturated temperature when air is present in the system. As shown in the following table, the temperature of the steam decreases as the percentage of air increases. A thermostatic air vent is specifically designed for removing air from a steam system.

In some special applications, it is necessary to use an air eliminator that will close when the vent body contains steam or water and opens when it contains air or gases. Combination float and thermostatic air vents are used in these applications.

Effects of Air on Temperature (°F) of a Steam and Air Mixture

Mixture Pressure psig	Pure Steam	5% Air	10% Air	15% Air
2	219°	216°	213°	210°
5	227°	225°	222°	219°
10	239°	237°	233°	230°
20	259°	256°	252°	249°

(Courtesy Spirax Sarco Co.)

HYDRONIC (HOT-WATER) HEATING SYSTEMS

Sizing Hydronic (Hot-Water) Boilers

To size a hydronic (hot-water) boiler, you will first need to make a heat-loss calculation for the structure. Then refer to the literature for the make and model of the boiler, and find the figure in the I=B=R Net Rating column that most closely matches the calculated heat loss. If the figure for the calculated heat loss is in between the closest matching I=B=R Net Rating figures, always choose the higher figure.

Sizes of Hot-Water Mains

Radiation (ft^2)	*Pipe (in)*
75 to 125	1¼
125 to 175	1½
175 to 300	2
300 to 475	2½
475 to 700	3
700 to 950	3½
950 to 1200	4
1200 to 1575	4½
1575 to 1975	5
1975 to 2375	5½
2375 to 2850	6

Residential and Light-Commercial Circulator Specifications

Taco Series 110–120 Red Baron Circulators

This series is designed to efficiently circulate heated or chilled water in residential or light commercial hydronic systems.

These circulators may also be used for zoning large installations and are available in bronze construction for domestic hot-water applications.

Electrical Data

Model	*Volts*	*Hertz*	*Phase*	*RPM*	*Horsepower*
110	115	60	1	1725	1/12
111	115	60	1	1725	1/8
112	115	60	1	3450	1/3
113	115	60	1	1725	1/8
120	115	60	1	1725	1/6

Motor Data

Motor type	Split-phase motor with built-in overload protection
Motor characteristics	Three-piece, mechanical seal-type circulator
Motor options	220/50/1, 220/60/1, 230/60/1, 100/110/50/60/1

Performance Data

Flow range	See the following performance chart
Head range	See the following performance chart
Maximum fluid temperature	240°F* (115°C)
Maximum working pressure	125 psi
Connection (flange) sizes	¾, 1, 1¼, 1½ in. (Models 110, 111, 112, 113); 2 in. (Model 120)

240°F intermittent; 200°F continuous

Taco Series 110–120 Red Baron Circulators *(continued)*

Performance Chart

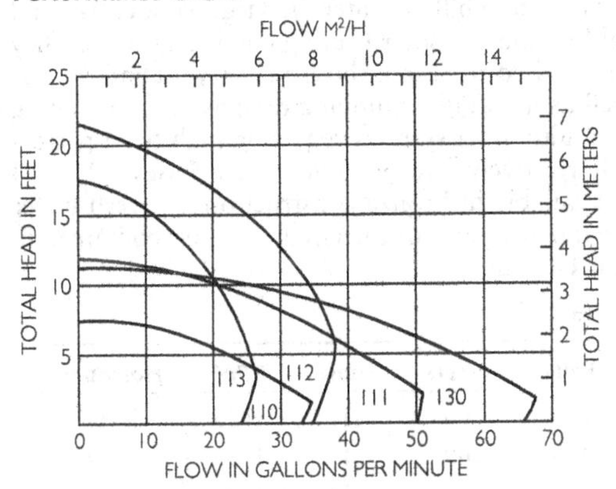

Pump Dimensions

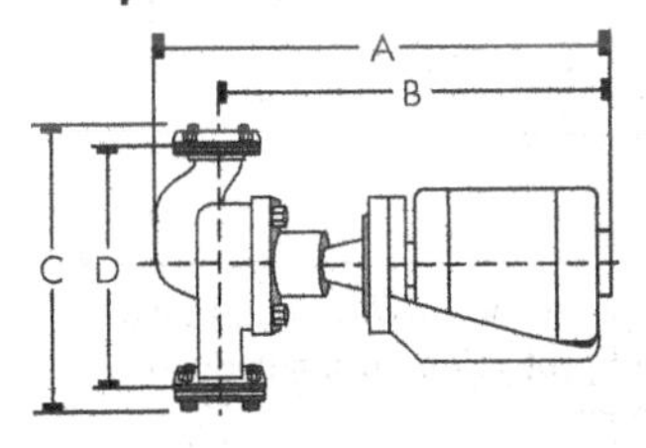

	A		B		C		D	
Model	***in.***	***mm***	***in.***	***mm***	***in.***	***mm***	***in.***	***mm***
110	14⅝	371.5	12⅝	320.7	7⅞	200	65/16	160.3
111	16¼	412.8	13⅞	352.4	10¼	260.4	8¾	222.3
112	16½	419.1	14½	368.3	7⅞	200	6⅜	161.9
113	16¼	412.8	14	355.6	10⅛	257.2	8½	215.9
120	16⅞	428.6	14¼	362	13½	342.9	11	279.4

Taco Series 121–133 Inline Cartridge Pumps

The Taco Series 121–133 Inline Cartridge Pumps are used to circulate heated or chilled water in large residential and commercial hydronic or solar heating/cooling systems. They are especially suited for use in both primary-secondary systems, as well as in parallel pumping designs, and in heating and/or air conditioning systems requiring high head performance with relatively low volume. Taco Series 121–133 pumps are available in bronze construction for fresh-water service. The moving parts of the circulators are contained in a replaceable cartridge.

Electrical Data

Model	*Volts*	*Hertz*	*Phase*	*RPM*	*Horsepower*
121	115	60	1	1725	¼
122	115	60	1	1725	¼
131	115	60	1	1725	¼
132	115*	60	1	1725	¾
133	115	60	1	1725	¾

** Motor options include 230/460/60/3 and 200/60/3*

Motor Data

Motor type	Split-phase motor with built-in overload protection on single-phase models.
Motor characteristics	Mechanical seal-type circulator, dipstick for checking oil level, drain plug for changing oil.
Motor options	220/50/1, 220/60/1, 230/60/1, 100/110/50/60/1

Taco Series 121–133 Inline Cartridge Pumps *(continued)*

Performance Data

Flow range	See Performance Chart
Head range	See Performance Chart
Maximum fluid temperature	240°F* (115°C)
Maximum working pressure	125 psi
Connection (flange) sizes	¾, 1, 1¼, 1½ in. (Models 110, 111, 112, 113); 2 in. (Model 120)

**240°F intermittent; 200°F continuous*

Performance Chart

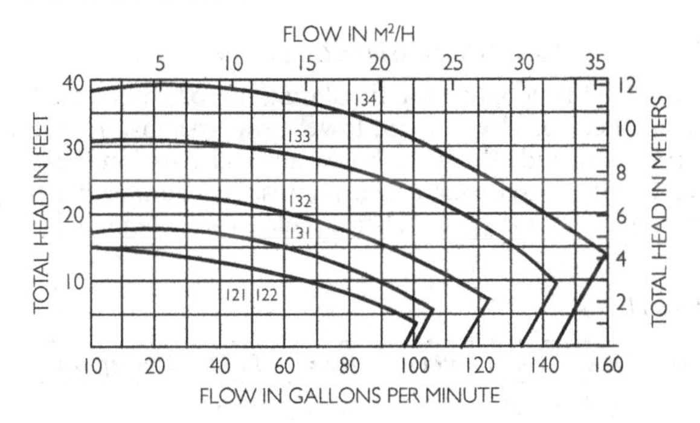

Pump Dimensions

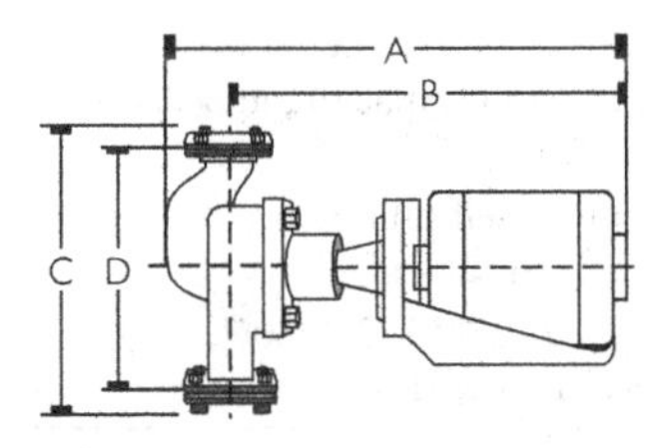

Taco Series 121–133 Inline Cartridge Pumps (continued)

	A		B		C		D	
Model	*in.*	*mm*	*in.*	*mm*	*in.*	*mm*	*in.*	*mm*
121	18⅛	460.4	15⅞	403.2	14¼	362	11⅛	282.6
122	18⅛	460.4	15⅞	403.2	13⅝	346.1	11⅛	282.6
131	19¼	489	15¾	400.1	16	406.4	13⅝	346.1
132	21½	546.1	18	457.2	16	406.4	13⅝	346.1
133	22⅛	562	18⅝	473.1	16	406.4	13⅝	346.1

Taco 003 Cartridge Circulator

The Taco 003 is designed for circulating hot or chilled fresh water in open- or closed-loop, lower-flow systems. Typical applications include domestic hot-water recirculation, heat-recovery units, water-source heat pumps, and potable water systems. The moving parts of the circulator are contained in a replaceable cartridge.

Electrical Data

Model	*Volts*	*Hertz*	*Phase*	*Amps*	*RPM*	*Horsepower*
003	115	60	1	0.43	3250	1/40

Motor Data

Motor type	Permanent split capacitor, impedance protected
Motor characteristics	Direct drive, self-lubricating, no mechanical seal
Motor options	220/50/1, 220/60/1, 230/60/1, 100/110/50/60/1

Taco 003 Cartridge Circulator *(continued)*

Performance Data

Flow range	0–6 gpm
Head range	0–5 ft
Minimum fluid temperature	40°F (4°C)
Maximum fluid temperature	220°F (104°C)
Maximum working pressure	125 psi
Connection sizes	½-in. Swt, ¾-in. Swt, ¾-in. NPT, or Union

Connection Data

Model	*Connection*
003-B4	¾ in. Swt
003-BC4	½ in. Swt
003-BT4	¾ in. NPT
003-BC4-1	Union

B = Bronze, Sweat
BC = Bronze, Sweat, Panel Mount Tappings
BT = Bronze, Threaded
BC-1 = Bronze, Union, Panel Mount Tappings

Taco 003 Cartridge Circulator *(continued)*

Pump Dimensions

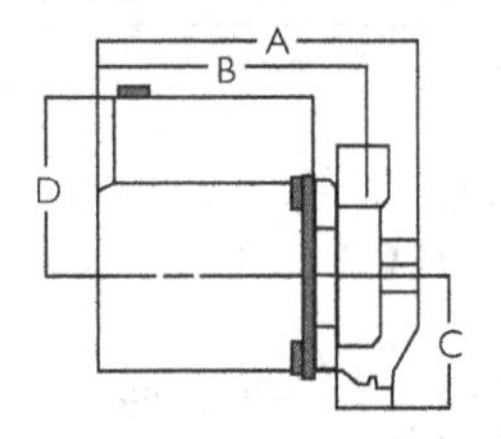

	A		B		C		D	
Model	*in.*	*mm*	*in.*	*mm*	*in.*	*mm*	*in.*	*mm*
003-B4	5⅛	130	4⅛	105	2³⁄₁₆	56	3¹⁄₁₆	78
003-BC4	5⅛	130	4⅛	105	2⅛	54	3¹⁄₁₆	78
003-BT4	5⅛	143	4⅞	124	2	51	3¹⁄₁₆	78
003-BC4-1	5⁵⁄₃₂	131	4¹¹⁄₃₂	110	2³¹⁄₃₂	76	3¹⁄₁₆	78

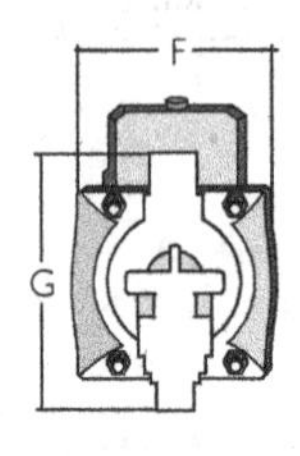

	F		G	
Model	*in.*	*mm*	*in.*	*mm*
003-B4	3⁵⁄₁₆	84	4¹³⁄₃₂	112
003-BC4	3⁵⁄₁₆	84	4¼	108
003-BT4	3⁵⁄₁₆	84	4	102
003-BC4-1	3⁵⁄₁₆	84	5¹⁵⁄₁₆	151

Taco 003 Cartridge Circulator *(continued)*

Mounting Positions

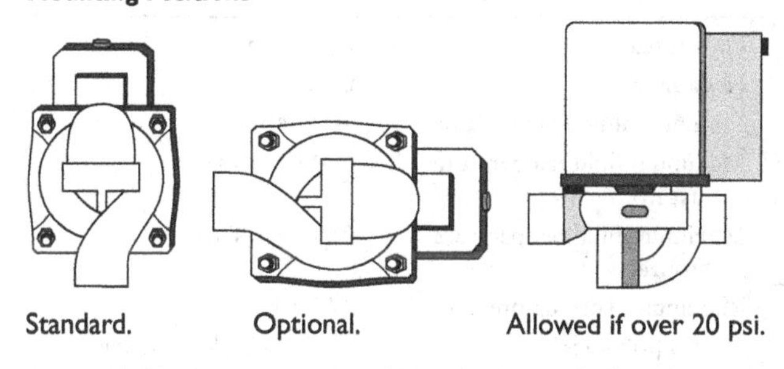

Standard. Optional. Allowed if over 20 psi.

Taco 005 Cartridge Circulator

The Taco 005 is designed for a wide range of residential and light-commercial water circulating applications. Typical applications include hydronic heating, add-a-zone additions, zoning with circulators, chilled water and domestic water systems. The moving parts of the circulator are contained in a replaceable cartridge.

Electrical Data

Model	*Volts*	*Hertz*	*Phase*	*Amps*	*RPM*	*Horsepower*
005-F2	115	60	1	0.53	3250	1/35
005-BF2	115	60	1	0.54	3250	1/35

Motor Data

Motor type	Permanent split capacitor, impedance protected
Motor characteristics	Direct drive, self-lubricating, no mechanical seal
Motor options	220/50/1, 220/60/1, 230/60/1, 100/110/50/60/1

Taco 005 Cartridge Circulator *(continued)*

Performance Data

Flow range	0–18 gpm
Head range	0–9.5 feet
Minimum fluid temperature	40°F (4°C)
Maximum fluid temperature (cast iron)	230°F (110°C)
Maximum fluid temperature (bronze)	220°F (104°C)
Maximum working pressure	125 psi
Connection sizes	¾, 1, 1¼, 1½ in. flanged

Pump Dimensions

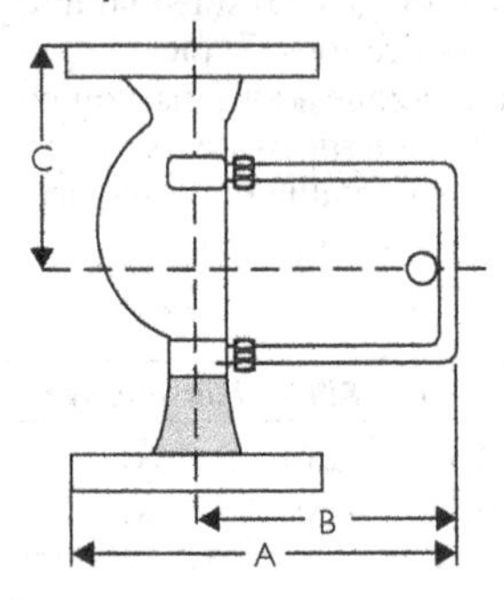

	A		B		C		D	
Model	**in.**	**mm**	**in.**	**mm**	**in.**	**mm**	**in.**	**mm**
005-F2	5⅝	143	4	102	3³⁄₁₆	81	2¹⁵⁄₁₆	75
005-BF2	5⅝	143	4	102	3³⁄₁₆	81	2¹⁵⁄₁₆	75

005-F2 = cast-iron construction
005-BF2 = bronze construction

Taco 005 Cartridge Circulator *(continued)*

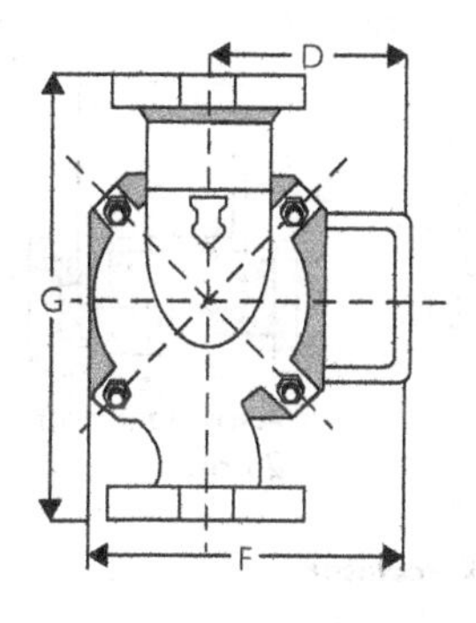

Model	F in.	F mm	G in.	G mm
005-F2	4¾	121	6⅜	162
005-BF2	4¾	121	6⅜	162

Flange Orientation

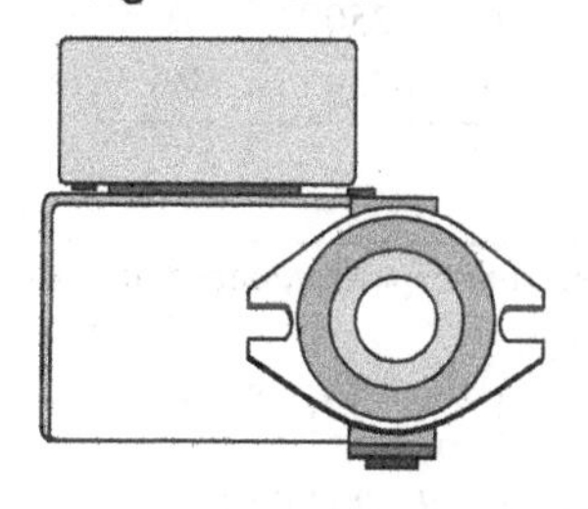

Taco 005 Cartridge Circulator *(continued)*

Mounting Positions

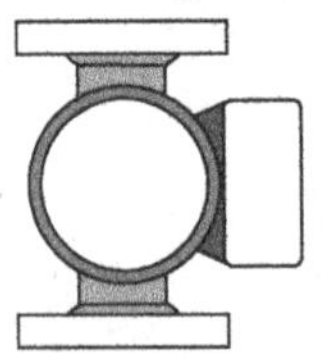

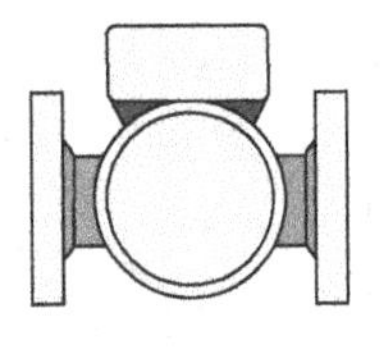

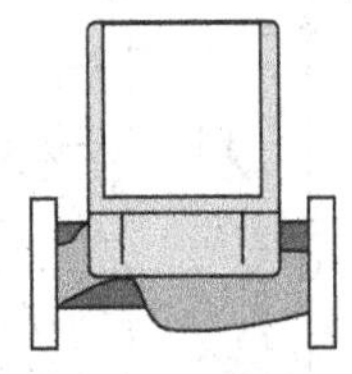

Standard. Optional. Allowed if over 29 psi.

Taco 007 Cartridge Circulator

The Taco 007 Cartridge Circulator is a wet-rotor, in-line, single-stage circulator pump. Typical applications include hydronic heating, radiant heating, hydro-air fan coils, indirect water heating, chilled fresh water, and domestic water systems. The moving parts of the circulator are contained in a replaceable cartridge.

Electrical Data

Model	*Volts*	*Hertz*	*Phase*	*Amps*	*RPM*	*Horsepower*
007-F5	115	60	1	0.70	3250	1/25
007-BF5	115	60	1	0.76	3250	1/25

Motor Data

Motor type	Permanent split capacitor, impedance protected
Motor characteristics	Direct drive, self-lubricating, no mechanical seal
Motor options	220/50/1, 220/60/1, 230/60/1, 100/110/50/60/1

Taco 007 Cartridge Circulator *(continued)*

Performance Data

Flow range	0–20 gpm
Head range	0–11 feet
Minimum fluid temperature	40°F (4°C)
Maximum fluid temperature (cast iron)	240°F (115°C)
Maximum fluid temperature (bronze)	230°F (110°C)
Maximum working pressure	125 psi
Minimum required inlet pressure	14 psi
Connection sizes	¾, 1, 1¼, 1½ in. flanged

Pump Dimensions

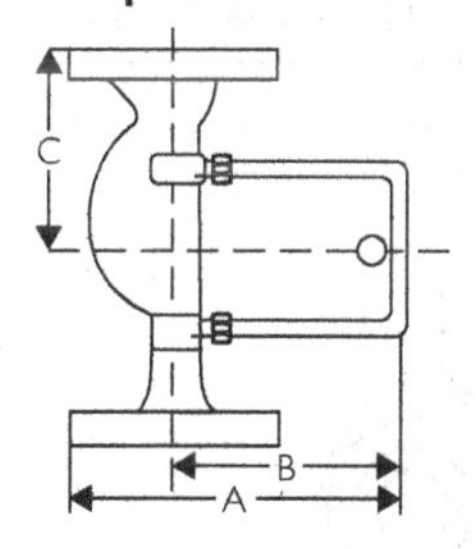

	A		B		C	
Model	***in.***	***mm***	***in.***	***mm***	***in.***	***mm***
007-F5	6⅛	156	4½	114	3³⁄₁₆	81
007-BF5	6⅛	156	4½	114	3³⁄₁₆	81

Taco 007 Cartridge Circulator *(continued)*

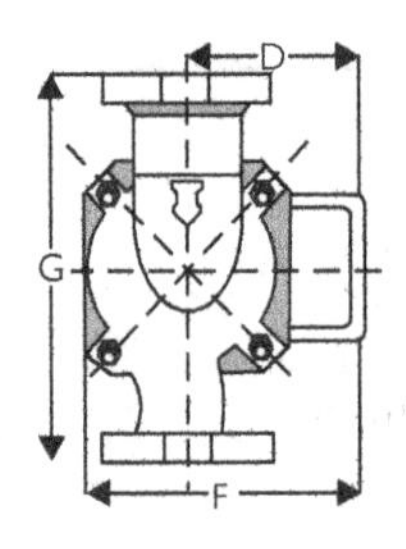

	D		F		G	
Model	***in.***	***mm***	***in.***	***mm***	***in.***	***mm***
007-F5	$2\frac{15}{16}$	75	$4\frac{3}{4}$	121	$6\frac{3}{8}$	162
007-BF5	$2\frac{15}{16}$	75	$4\frac{3}{4}$	121	$6\frac{3}{8}$	162

Flange Orientation

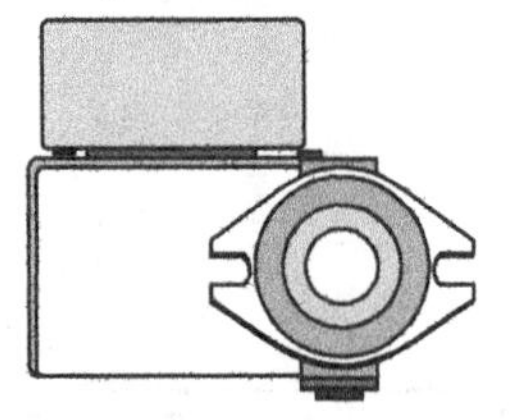

Standard.

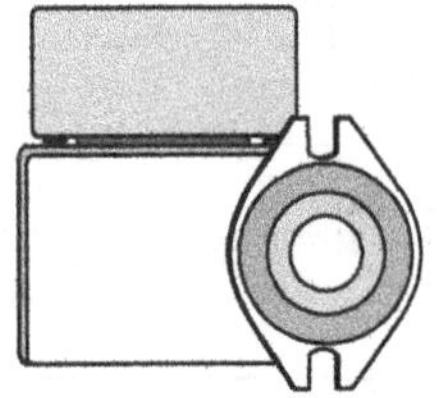
Optional.

Taco 007 Cartridge Circulator *(continued)*

Mounting Positions

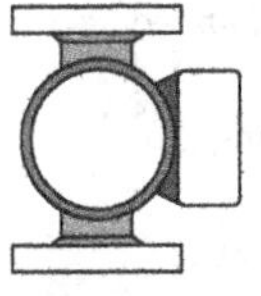

Standard.

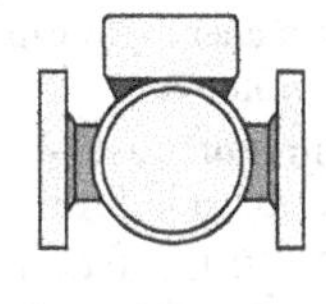

Optional.

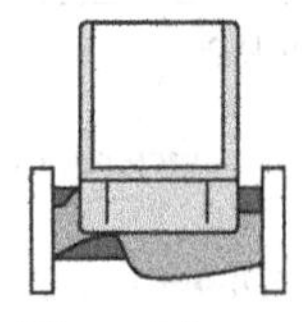

Allowed if over 20 psi.

Taco 007-IFC Cartridge Circulator

The Taco 007-IFC Cartridge Circulator is designed with an integral flow-check valve and a replaceable cartridge. The moving parts of the circulator are contained in the replaceable cartridge. The Taco Integral Flow Check (IFC) eliminates the necessity for a separate in-line flow check valve. The reduced pressure drop of the IFC increases the 007 flow performance up to 240 percent over in-line check valves. Typical applications include hydronic or radiant heating, hydro-air fan coils, or indirect water heaters.

The Taco 0010-Z-IFC combines the circulator, flow-check and a priority zoning control with built-in transformer, relay, and priority switch in one complete unit for zoning applications. The Taco 0010-IFC is available in both cast-iron and bronze construction.

Electrical Data

Model	*Volts*	*Hertz*	*Phase*	*Amps*	*RPM*	*Horsepower*
007-IFC (cast iron)	115	60	1	0.71	3250	1/25
007-IFC (bronze)	115	60	1	0.72	3250	1/25

Taco 007-IFC Cartridge Circulator *(continued)*

Motor Data

Motor type	Permanent split capacitor, impedance protected
Motor characteristics	Direct drive, self-lubricating, no mechanical seal
Motor options	220/50/1, 220/60/1, 230/60/1, 100/110/50/60/1

Performance Data

Flow range	0–12.5 gpm
Head range	0–11 feet
Minimum fluid temperature	40°F (4°C)
Maximum fluid temperature	240°F (115°C)
Maximum working pressure	125 psi
Connection sizes	¾, 1, 1¼, 1½ in. flanged

Pump Dimensions

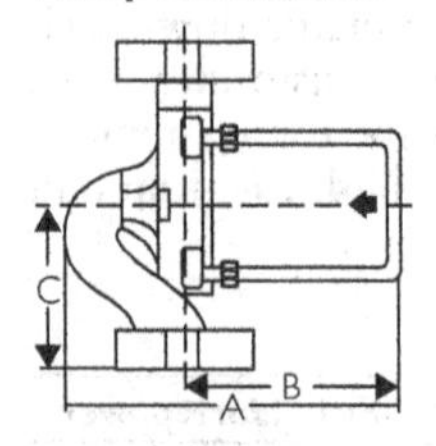

Taco 007-IFC Cartridge Circulator *(continued)*

Model	A		B		C	
	in.	*mm*	*in.*	*mm*	*in.*	*mm*
007-F5-IFC	7	178	4½	114	3 3/16	81
007-BF5-IFC	7	178	4½	114	3 3/16	81
007-ZF5-IFC	7	178	4½	114	3 3/16	81
007-ZBF5-IFC	7	178	4½	114	3 3/16	81

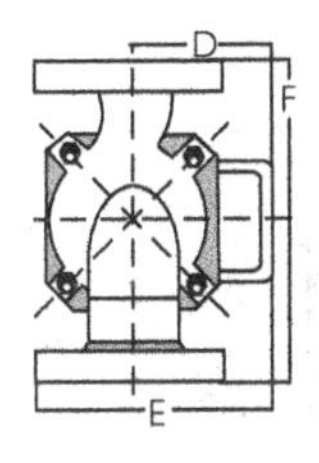

Model	D		E		F	
	in.	*mm*	*in.*	*mm*	*in.*	*mm*
007-F5-IFC	2 15/16	75	5	127	6 3/8	162
007-BF5-IFC	2 15/16	75	5	127	6 3/8	162
007-ZF5-IFC	3 5/8	92	5 9/16	143	6 3/8	162
007-ZBF5-IFC	3 5/8	92	5 9/16	143	6 3/8	162

Taco 007-IFC Cartridge Circulator *(continued)*

Flange Orientation

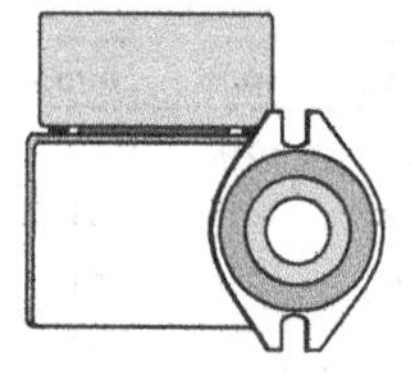

Mounting Positions

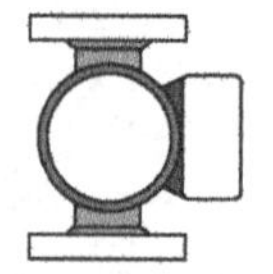

Standard.

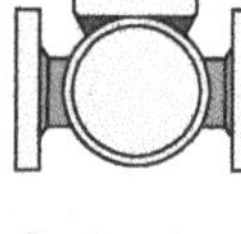

Optional.

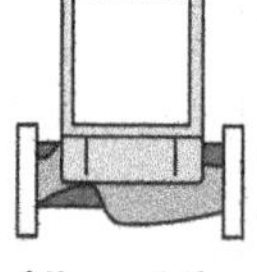

Allowed if over 20 psi.

Taco 0010 Cartridge Circulator

The Taco 0010 Cartridge Circulator is designed for a wide range of large-residential and light-commercial water-circulating applications. Typical applications include hydronic heating, radiant heating, primary-secondary loops, indirect water heaters, chilled water cooling, and potable hot-water systems. The Taco 0010 (BF2 bronze casing) is designed for all open-loop, fresh water systems. A replaceable cartridge on the Taco 0010 contains the moving parts of the circulator.

Electrical Data

Model	*Volts*	*Hertz*	*Phase*	*Amps*	*RPM*	*Horsepower*
0010-F2	115	60	1	1.10	3250	⅛
0010-BF2	115	60	1	1.17	3250	⅛

Taco 0010 Cartridge Circulator *(continued)*

Motor Data

Motor type	Permanent split capacitor, impedance protected
Motor characteristics	Direct drive, self-lubricating, no mechanical seal
Motor options	220/50/1, 220/60/1, 230/60/1, 100/110/50/60/1

Performance Data

Flow range	0–30 gpm
Head range	0–11.5 ft
Minimum fluid temperature	40°F (4°C)
Maximum fluid temperature	230°F (110°C)
Maximum working pressure	125 psi
Connection sizes	¾, 1, 1¼, 1½ in. flanged

Pump Dimensions

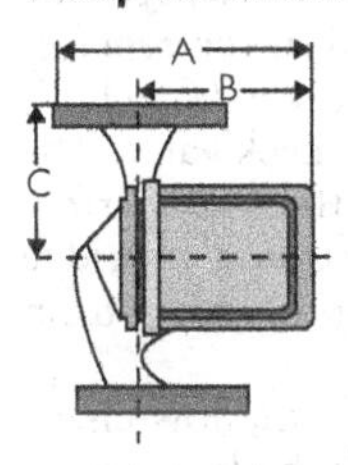

	A		B		C	
Model	***in.***	***mm***	***in.***	***mm***	***in.***	***mm***
0010-F2	7¼	184	5 5/16	135	3 3/16	81
0010-BF2	7¼	184	5 5/16	135	3 3/16	81

Taco 0010 Cartridge Circulator *(continued)*

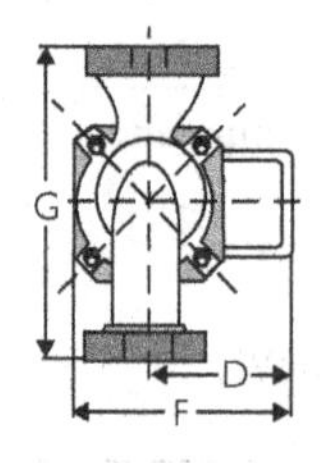

Model	D		F		G	
	in.	*mm*	*in.*	*mm*	*in.*	*mm*
0010-F2	3 5/16	84	5 3/8	137	6 3/8	162
0010-BF2	3 5/16	84	5 3/8	137	6 3/8	162

Taco 0010-IFC Cartridge Circulator

The Taco 0010-IFC Cartridge Circulator is designed with an integral flow-check valve and a replaceable cartridge. The moving parts of the circulator are contained in the replaceable cartridge. The Taco Integral Flow Check (IFC) eliminates the necessity for a separate in-line flow check valve. The reduced pressure drop of the IFC increases the flow performance over in-line check valves. Both the IFC and cartridge can be accessed for service without removing the pump from the system.

The Taco 0010-Z-IFC combines the circulator, flow-check and patented Priority Zoning control with built-in transformer, relay, and priority switch in one complete unit for zoning applications. The Taco 0010-IFC is available in both cast-iron and bronze construction.

Taco 0010-IFC Cartridge Circulator *(continued)*

Electrical Data

Model	*Volts*	*Hertz*	*Phase*	*Amps*	*RPM*	*Horsepower*
0010-F3-IFC	115	60	1	1.10	3250	⅛
0010-BF3-IFC	115	60	1	1.17	3250	⅛
0010-ZF3-IFC	115	60	1	1.10	3250	⅛
0010-ZBF3-IFC	115	60	1	1.17	3250	⅛

Motor Data

Motor type	Permanent split capacitor, impedance protected
Motor characteristics	Direct drive, self-lubricating, no mechanical seal
Motor options	220/50/1, 220/60/1, 230/60/1, 100/110/50/60/1

Performance Data

Flow range	0–18.5 gpm
Head range	0–9 ft
Minimum fluid temperature	40°F (4°C)
Maximum fluid temperature	230°F (110°C)
Maximum working pressure	125 psi
Connection sizes	¾, 1, 1¼, 1½ in. flanged

Taco 0010-IFC Cartridge Circulator *(continued)*

Pump Dimensions

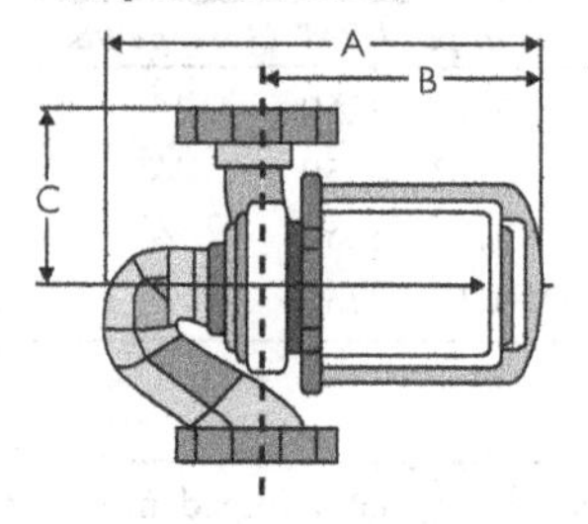

Model	*Casing*	**A**		**B**		**C**	
		in.	*mm*	*in.*	*mm*	*in.*	*mm*
0010-F3-IFC	Cast iron	8⅜	213	5$^{5}/_{16}$	135	3$^{3}/_{16}$	81
0010-BF3-IFC	Bronze	8⅜	213	5$^{5}/_{16}$	135	3$^{3}/_{16}$	81
0010-ZF3-IFC	Cast iron	8⅜	213	5$^{5}/_{16}$	135	3$^{3}/_{16}$	81
0010-ZBF3-IFC	Bronze	8⅜	213	5$^{5}/_{16}$	135	3$^{3}/_{16}$	81

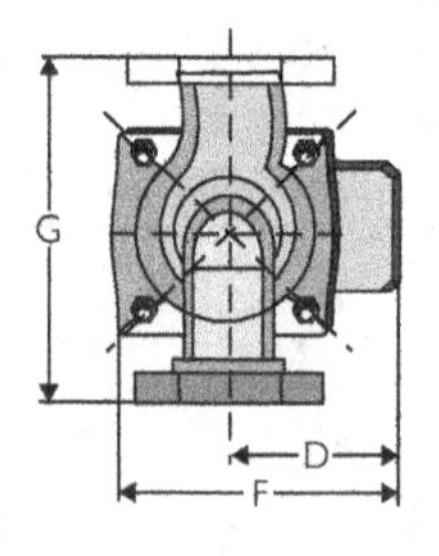

Taco 0010-IFC Cartridge Circulator *(continued)*

Model	*Casing*	*D*		*F*		*G*	
		in.	*mm*	*in.*	*mm*	*in.*	*mm*
0010-F3-IFC	Cast iron	3⁵⁄₁₆	84	5⅜	137	6⅜	162
0010-BF3-IFC	Bronze	3⁵⁄₁₆	84	5⅜	137	6⅜	162
0010-ZF3-IFC	Cast iron	3¹³⁄₁₆	97	5⅞	150	6⅜	162
0010-ZBF3-IFC	Bronze	3¹³⁄₁₆	97	5⅞	150	6⅜	162

Flange Orientation

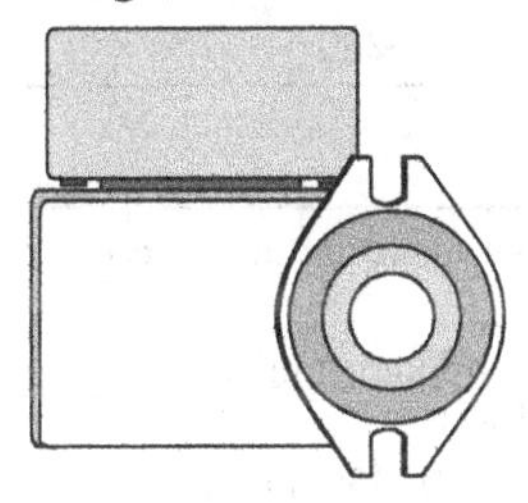

Mounting Positions

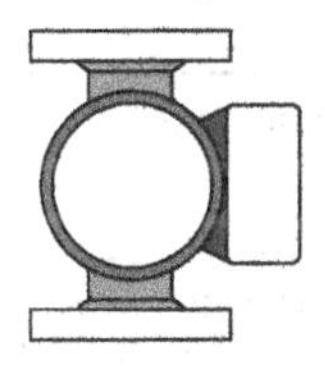

Standard.

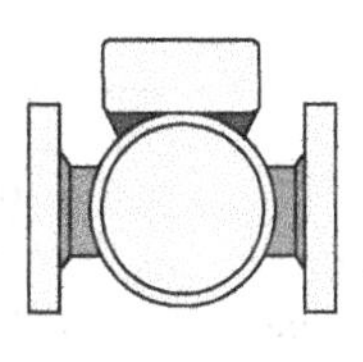

Optional.

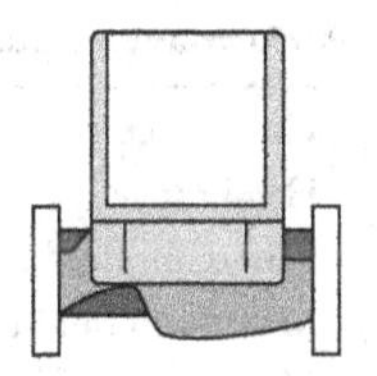

Allowed if over 20 psi.

Taco 0011 Cartridge Circulator

The Taco 0011 is designed for high-head/medium-flow applications in large residential/light-commercial closed-loop hydronic heating and chilled water cooling systems. Typical applications include high-pressure drop boilers, fan-coil units, heat exchangers, large radiant-heating and heat-recovery/geothermal systems. The Bronze 0011 can be used on open-loop systems. All the moving parts of the circulator are contained in a replaceable cartridge.

Electrical Data

Model	*Volts*	*Hertz*	*Phase*	*Amps*	*RPM*	*Horsepower*
0011-F4	115	60	1	1.76	3250	⅛
0011-BF4	115	60	1	1.76	3250	⅛

Motor Data

Motor type	Permanent split capacitor, impedance protected
Motor characteristics	Direct drive, self-lubricating, no mechanical seal
Motor options	220/50/1, 220/60/1, 230/60/1, 100/110/50/60/1

Performance Data

Flow range	0–28 gpm
Head range	0–30 ft
Minimum fluid temperature	40°F (4°C)
Maximum fluid temperature	230°F (110°C)
Maximum working pressure	125 psi
Minimum required inlet pressure	14 psi
Connection sizes	¾, 1, 1¼, 1½ in.

Taco 0011 Cartridge Circulator *(continued)*

Pump Dimensions

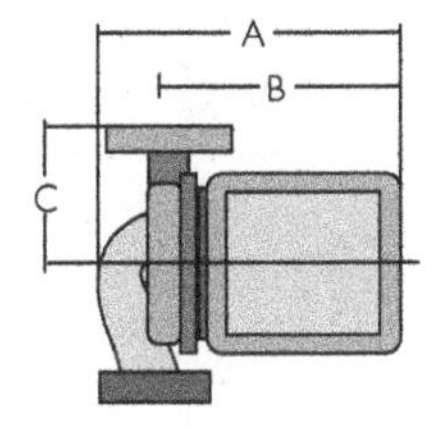

Model	A		B		C	
	in.	*mm*	*in.*	*mm*	*in.*	*mm*
0011-F4 (cast iron)	7⅝	194	6 7/32	158	3¼	82
0011-BF4 (bronze)	7⅝	194	6 7/32	158	3¼	82

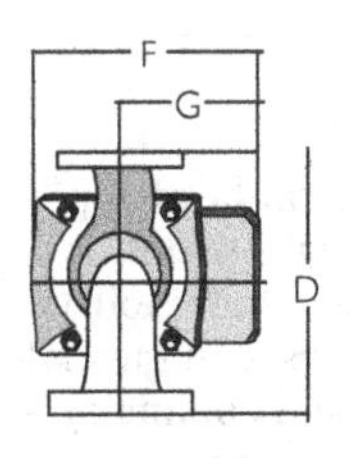

Model	D		F		G	
	in.	*mm*	*in.*	*mm*	*in.*	*mm*
0011-F4 (cast iron)	3¾	95	6	152	6½	165
0011-BF4 (bronze)	3¾	95	6	152	6½	165

Taco 0011 Cartridge Circulator *(continued)*

Flange Orientation

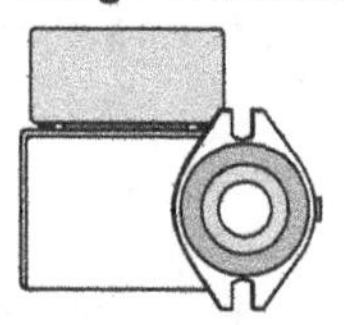

Mounting Positions

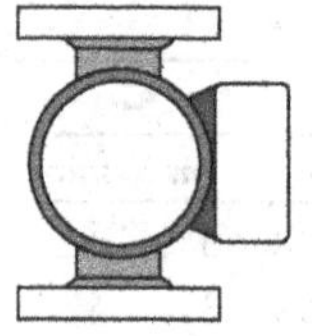

Standard.

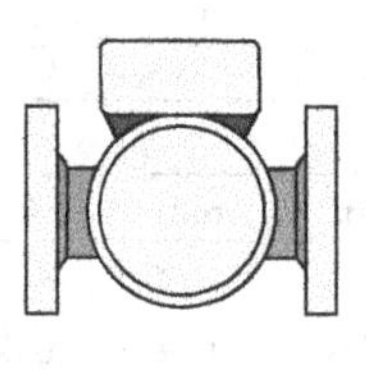

Optional.

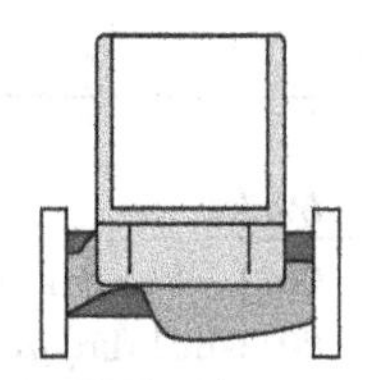

Allowed if over 20 psi.

Taco 0012 Cartridge Circulator

The Taco 0012 is designed for high-flow/medium-head applications in large-residential and light-commercial systems. It is commonly used in conjunction with Ideal large Btu/h Boilers, 2" recirculation loops, primary/secondary loops, commercial water heaters and light-commercial heating and cooling systems. The Bronze 0012 can be used on open-loop systems. The replaceable cartridge contains all the moving parts of the circulator. Universal flange-to-flange dimensions and orientation allows the 0012 to easily replace other models. The Taco 0012-F4 circulator is a direct replacement for the HV Series, and the 0012-F4-1 is a direct replacement for the 2 inch Series using existing flanges.

Taco 0012 Cartridge Circulator *(continued)*

Electrical Data

Model	*Volts*	*Hertz*	*Phase*	*Amps*	*RPM*	*Horsepower*
0012	115	60	1	1.33	3250	⅛

Motor Data

Motor type	Permanent split capacitor, impedance protected
Motor characteristics	Direct drive, self-lubricating, no mechanical seal
Motor options	220/50/1, 220/60/1, 230/60/1, 100/110/50/60/1

Performance Data

Flow range	0–50 gpm
Head range	0–14.5 ft
Minimum fluid temperature	40°F (4°C)
Maximum fluid temperature	230°F (110°C)
Maximum working pressure	125 psi
Minimum Required inlet pressure	14 psi
Connection sizes	1¼-in., 1½-in. flanged or 2-in. flanged

Taco 0012 Cartridge Circulator *(continued)*

Pump Dimensions

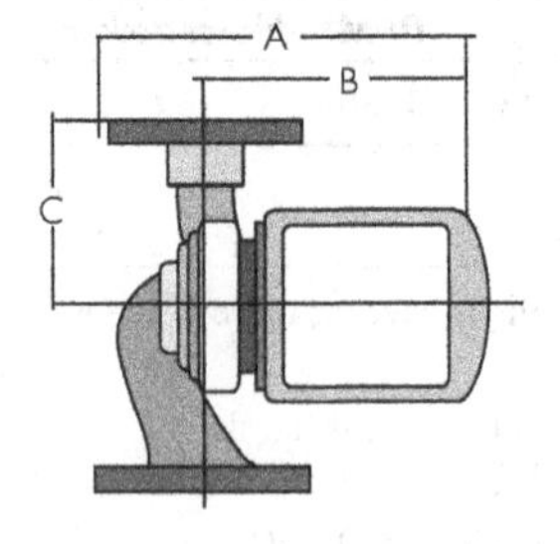

	A		***B***		***C***	
Model	***in.***	***mm***	***in.***	***mm***	***in.***	***mm***
0012-F4	8⅝	219	6⅜	162	4¼	108
0012-F4-1	9	229	6⅜	162	4¼	108
0012-BF4	8⅝	219	6⅜	162	4¼	108
0012-BF4-1	9	229	6⅜	162	4¼	108

Model	***Casing***	***Flange***
0012-F4	Cast iron	1¼", 1½"
0012-F4-1	Cast iron	2"
0012-BF4	Bronze	1¼", 1½"
0012-BF4-1	Bronze	2"

Taco 0012 Cartridge Circulator *(continued)*

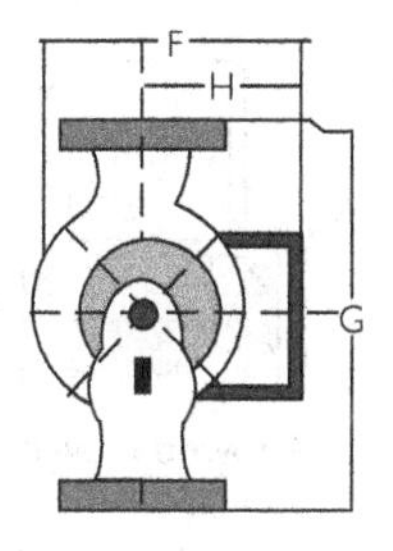

Model	D in.	D mm	F in.	F mm	G in.	G mm
0012-F4	3⅞	98	6	152	8½	216
0012-F4-1	3⅞	98	6	152	8½	216
0012-BF4	3⅞	98	6	152	8½	216
0012-BF4-1	3⅞	98	6	152	8½	216

Flange Orientation

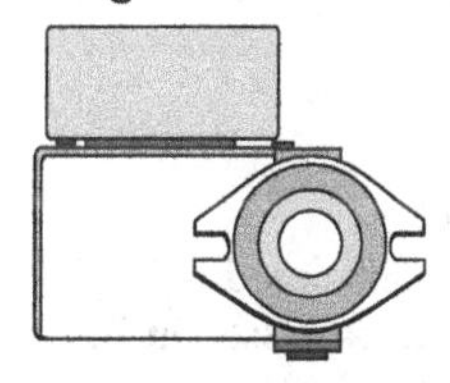

Taco 0012 Cartridge Circulator *(continued)*

Mounting Positions

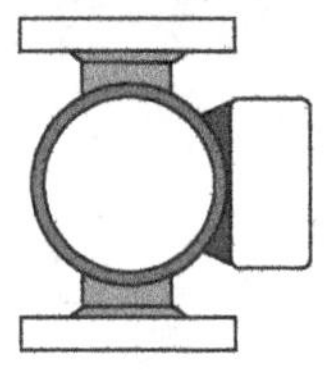

Standard.

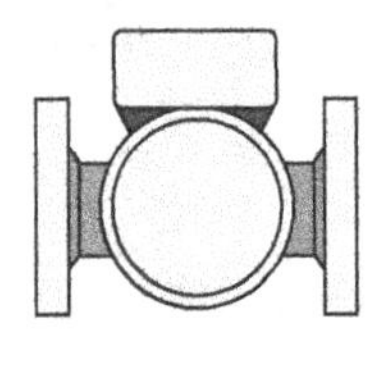

Optional.

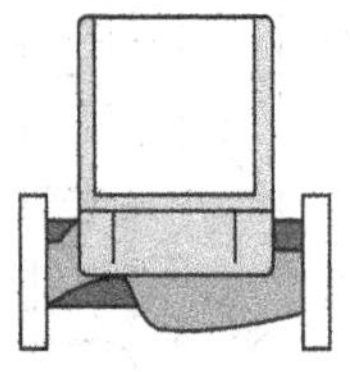

Allowed if over 20 psi.

Sizing Circulators (Centrifugal Pumps)

A rule-of-thumb method for sizing a circulator (water pump) for a hydronic heating system requires the following information:

- The estimated total heating load for the structure (expressed in Btu/h)
- The design temperature drop
- The required circulation rate (gpm) of the water.

$$\text{gpm} = \frac{\text{total heating load (in Btu/h)}}{T \times 60 \times 8}$$

Where

gpm	= gallons per minute
Total heating load	= total heating load calculated for the structure and expressed in Btu per hour (Btu/h)
T	= design temperature drop
60	= minutes per hour
8	= weight (in pounds) of a gallon of water

Assuming a design temperature drop (T) of 20°F (commonly used for most hydronic floor heating systems) and a total heating load of 30,000 Btu/h for the structure, the required rate of water circulation is as follows:

$$\text{gpm} = \frac{\text{total heating load}}{20 \times 60 \times 8}$$

$$= \frac{30{,}000 \text{ Btu/h}}{9600}$$

$$= 3.13$$

The hydronic heating system used in this example would require a circulator capable of moving the water at a rate of 3.13 gallons per minute.

Installing Expansion Tanks

An undersized expansion tank, or one that is completely filled up with water, will cause the boiler pressure to increase when the water heats. Because the expansion tank is too small or too filled with water to absorb the excess pressure, the relief valve will begin to drip. The dripping relief valve is only symptomatic of the real problem, and replacing the valve will in no way solve it.

There is not much you can do about an undersized expansion tank except replace it. As a rule-of-thumb, expansion tanks should be sized at 1 gallon for every 23 square feet of radiation, or 1 gallon for every 3500 Btu of radiation installed on the job.

If the problem is a completely filled tank, the tank should be partially drained so that there is enough space to permit future expansion under pressure. The first step in draining an expansion tank is to open the drain valve. The water will gush out at first in a heavy flow and then tend to gurgle out because a vacuum is building up inside the tank. Inserting a tube into the drain valve opening will admit air and break the vacuum, and the water will return to its normal rate of flow. After a sufficient amount of water has been removed, the drain valve can be closed.

Recommended Sizes for Expansion Tanks

Open System

Nominal Capacity—Gallons	***Square Feet of Radiation***
10	300
15	500
20	700
26	950

Closed System

Nominal Capacity—Gallons	***Square Feet of Radiation***
18	350
21	450
24	650
30	900
35	900
35	1100

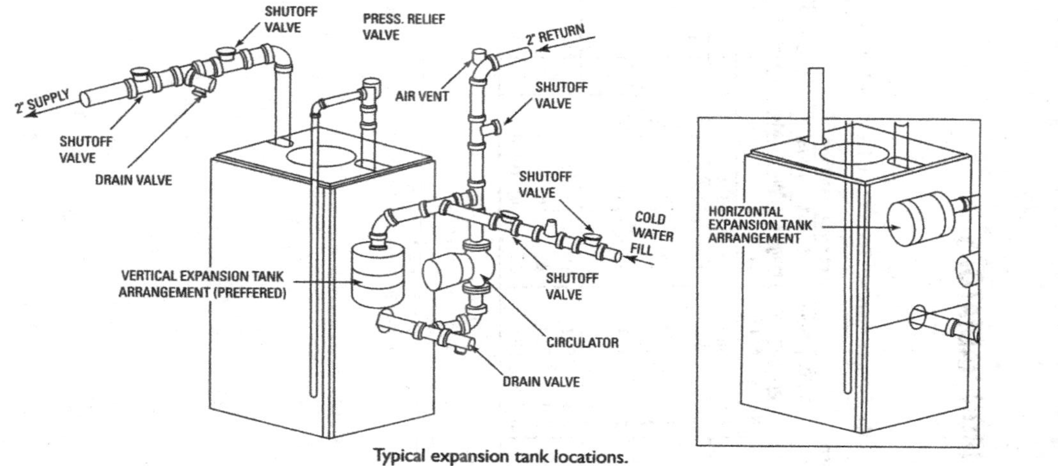

Typical expansion tank locations.

CAST-IRON RADIATORS

Ratings for Small-Tube Cast-Iron Radiators

Number of Tubes per Section	Section Dimensions						
	Catalog Rating per Section*		A Height‡	B Width		C Spacing†	D Leg Height‡
				Min	Max		
	ft^2	Btu/h	in	in	in	in	in
3§	1.6	384	25	3¼	3½	1¾	2½
4§	1.6	384	19	4 7/16	4 13/16	1¾	2½
	1.8	432	22	4 7/16	4 13/16	1¾	2½
	2.0	480	25	4 7/16	4 13/16	1¾	2½
5§	2.1	504	22	5⅝	6 5/16	1¾	2½
	2.4	576	25	5⅝	6 5/16	1¾	2½
6§	2.3	552	19	6 13/16	8	1¾	2½
	3.0	720	25	6 13/16	8	1¾	2½
	3.7	888	32	6 13/16	8	1¾	2½

*These ratings are based on steam at 215°F and air at 70°F. They apply only to installed radiators exposed in a normal manner, not to radiators installed behind enclosures, behind grilles, or under shelves.

†Length equals number of sections times 1¾ in.

‡Overall height and leg height, as produced by some manufacturers, are 1 inch greater than shown in columns A and D. Radiators may be furnished without legs. Where greater than standard leg heights are required, this dimension shall be 4½ in.

§Or equal.

(Courtesy 1960 ASHRAE Guide)

Ratings for Large-Tube Cast-Iron Radiators (sectional, cast-iron, tubular-type radiators of the large-tube pattern, that is, having tubes approximately 1⅜ inches in diameter, 2½ inches on center)

Number of Tubes per Section	Catalog Rating per Section*		Height	Width	Section Center Spacing†	Leg Height‡ to Tapping
	ft²	Btu/h	in	in	in	in
3	1¾	420	20	4⅝	2½	4½
	2	480	23		2½	4½
	2⅓	560	26		2½	4½
	3	720	32		2½	4½
	3½	840	38		2½	4½
4	2¼	540	20			
	2½	600	23		2½	4½
	2¾	660	26	6¼–6¹³⁄₁₅	2½	4½
	3½	840	32		2½	4½
	4¼	1020	38		2½	4½
5	2⅔	640	20		2½§	4½
	3	720	23		2½§	4½
	3½	840	26	9–8⁹⁄₁₈	2½§	4½
	4⅓	1040	32		2½§	4½
	5	1200	38		2½§	4½
6	3	720	20		2½	4½
	3½	840	23		2½	4½
	4	960	26	9–10⅜	2½	4½
	5	1200	32		2½	4½
	6	1440	38		2½	4½
7	2½	600	14		2½	3
	3	720	17	11⅜–12¹³⁄₁₆	2½	3
	3⅔	880	20		2½	3 or 4½

*These ratings are based on steam at 215°F and air at 70°F. They apply only to installed radiators exposed in a normal manner, not to radiators installed behind enclosures, behind grilles, or under shelves.

†Maximum assembly 60 sections. Length equals number of sections times 2½ in.

‡Where greater than standard leg heights are required, this dimension shall be 6 in, except for 7-tube sections, in heights from 13 to 20 in, inclusive, for which this dimension shall be 4½ in. Radiators may be furnished without legs.

§For five-tube hospital-type radiation, this dimension is 3 in.

(Courtesy 1960 ASHRAE Guide)

Rating for Column-Type Cast-Iron Radiators

	*Generally Accepted Rating per Section**					
	One-Column		**Two-Column**		**Three-Column**	
Height (in)	**ft²**	**Btu/h**	**ft²**	**Btu/h**	**ft²**	**Btu/h**
15			1½	360		
18					2¼	540
20	1½	360	2	480		
22			2¼	540	3	720
23	1⅔	400	2⅓	560		
26	2	480	2⅔	640	3¾	900
32	2½	600	3⅓	800	4½	1080
38	3	720	4	960	5	1200
45			5	1200	6	1440

	Four-Column		**Five-Column**		**Six-Column**	
Height (in)	**ft²**	**Btu/h**	**ft²**	**Btu/h**	**ft²**	**Btu/h**
13					3	720
16					3¾	900
18	3	720	4⅔	1120	4½	1080
20					5	1200
22	4	960				
26	5	1200	7	1680		
32	6½	1560				
38	8	1920	10	2400		
45	10	2400				

**These ratings are based on steam at 215°F and air at 70°F. They apply only to installed radiators exposed in a normal manner, not to radiators installed behind enclosures, behind grilles, or under shelves.*

(Courtesy 1960 ASHRAE Guide)

Sizing Cast-Iron Radiators

To size a column-type or tube-type cast-iron radiator, first measure its height in inches and then count the number of sections and the number of tubes or columns in each section.

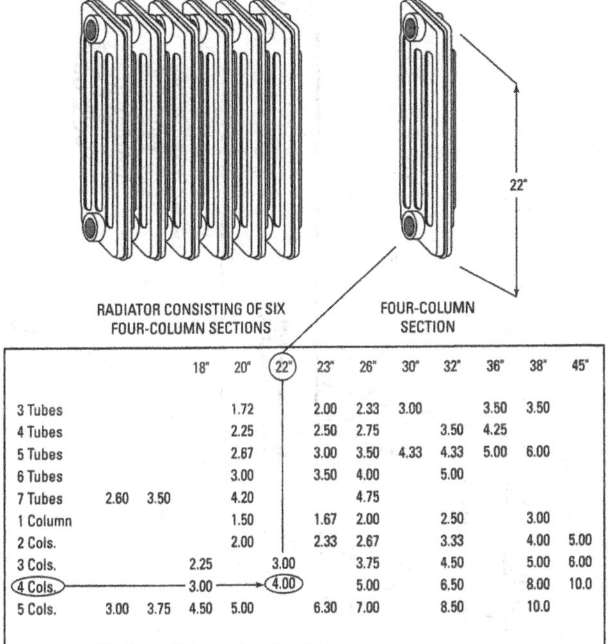

			18"	20"	22"	23"	26"	30"	32"	36"	38"	45"
3 Tubes				1.72		2.00	2.33	3.00		3.50	3.50	
4 Tubes				2.25		2.50	2.75		3.50	4.25		
5 Tubes				2.67		3.00	3.50	4.33	4.33	5.00	6.00	
6 Tubes				3.00		3.50	4.00		5.00			
7 Tubes	2.60	3.50		4.20			4.75					
1 Column				1.50		1.67	2.00		2.50		3.00	
2 Cols.				2.00		2.33	2.67		3.33		4.00	5.00
3 Cols.			2.25		3.00		3.75		4.50		5.00	6.00
4 Cols.			3.00		4.00		5.00		6.50		8.00	10.0
5 Cols.	3.00	3.75	4.50	5.00		6.30	7.00		8.50		10.0	

1. Find the Sq. Ft. EDR per section above (4.00).
2. Multiply by the number of sections altogether in that radiator to get Sq. Ft. EDR for the entire radiator ($4.00 \times 6 = 24$ Sq. Ft. EDR).
3. Multiply by 240 Btus per hour to get the steam design output or Multiple by 170 Btus per hour to get the hot water design output.

RADIATOR AND CONVECTOR DETAILS

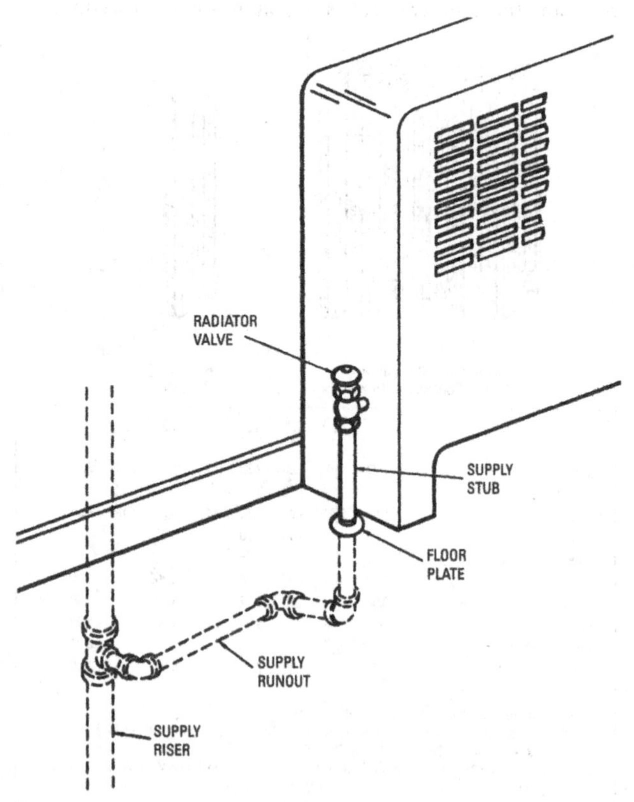

Steam convector showing supply runout location. *(Courtesy Dunham-Bush, Inc.)*

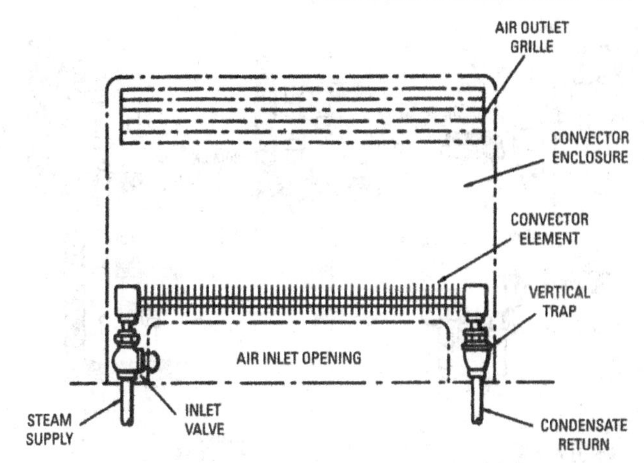

Steam convector connections to supply and condensate-return piping. *(Courtesy 1960 ASHRAE Guide)*

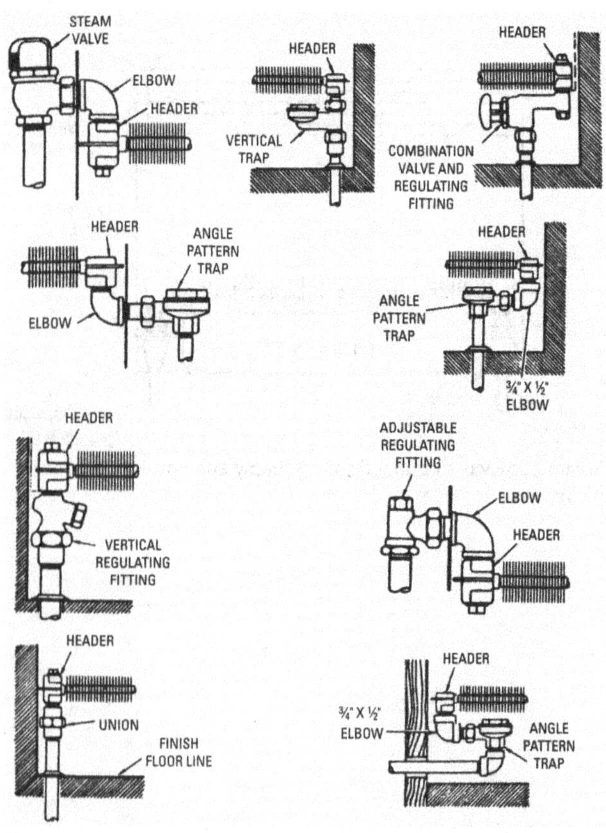

Convector piping connections in a steam heating system.

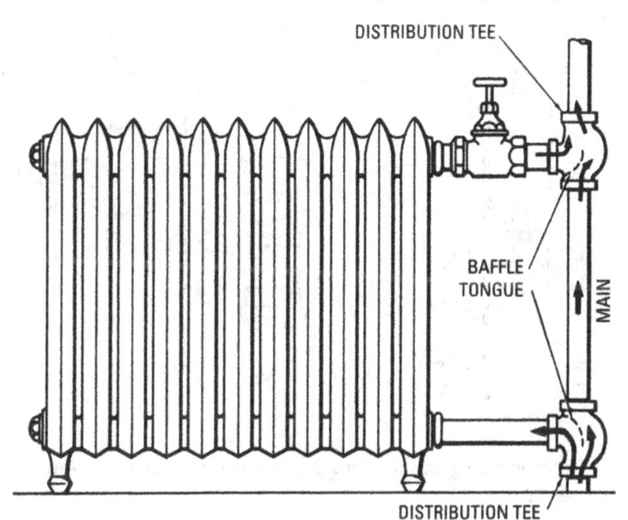

Using distribution tees to connect a cast-iron radiator to the main in a one-pipe hot-water heating system.

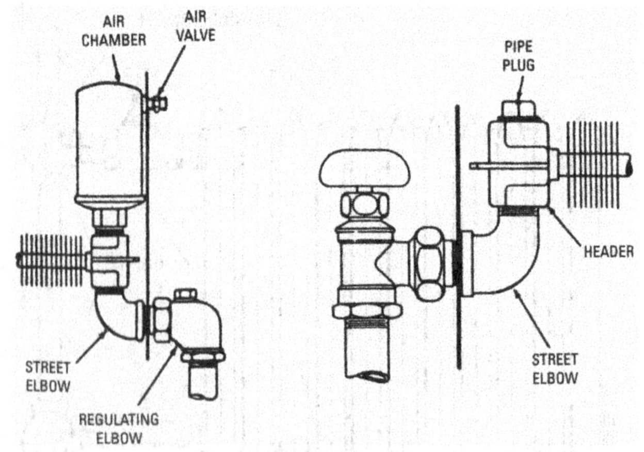

Convector piping connections in a hot-water heating system.
(Courtesy Dunham-Bush, Inc.)

PROPERTIES OF WATER, AIR, AND SATURATED STEAM

Properties of Dry Air

Temperature °F	*Weight per Cu Ft (F^3) of Dry Air in Pounds*	*Ratio to Volume at 70°F*	*Btu Absorbed per Cu Ft (F^3) of Air per °F*	*Cu Ft (F^3) of Air Raised 1°F by 1 Btu*
0	.08636	.8680	.02080	48.08
10	.08453	.8867	.02039	49.05
20	.08276	.9057	.01998	50.05
30	.08107	.9246	.01957	51.10
40	.07945	.9434	.01919	52.11
50	.07788	.9624	.01881	53.17
60	.07640	.9811	.01846	54.18
70	.07495	1.0000	.01812	55.19
80	.07356	1.0190	.01779	56.21
90	.07222	1.0380	.01747	57.25
100	.04093	1.0570	.01716	58.28
110	.06968	1.0756	.01687	59.28
120	.06848	1.0945	.01659	60.28
130	.06732	1.1133	.01631	61.32
140	.06620	1.1320	.01605	62.31
150	.06510	1.1512	.01578	63.37
160	.06406	1.1700	.01554	64.35
170	.06304	1.1890	.01530	65.36
180	.06205	1.2080	.01506	66.40

Weight of Water per Cubic Foot at Different Temperatures

Temp., °F	lb per ft³	Temp., °F	lb per ft³	Temp., °F	lb per ft³	Temp., °F	lb per ft³
32	62.41	55	62.38	78	62.23	101	61.98
33	62.41	56	62.38	79	62.22	102	61.96
34	62.42	57	62.38	80	62.21	103	61.95
35	62.42	58	62.37	81	62.20	104	61.94
36	62.42	59	62.37	82	62.19	105	61.93
37	62.42	60	32.36	83	62.18	106	61.91
38	62.42	61	62.35	84	62.17	107	61.90
39	62.42	62	62.35	85	62.16	108	61.89
40	62.42	63	62.34	86	62.15	109	61.87
41	62.42	64	62.34	87	62.14	110	61.86
42	62.42	65	62.33	88	62.13	111	61.84
43	62.42	66	62.32	89	62.12	112	61.83
44	62.42	67	62.32	90	62.11	113	61.81
45	62.42	68	62.31	91	62.10	114	61.80
46	62.41	69	62.30	92	62.08	115	61.78
47	62.41	70	62.30	93	62.07	116	61.77
48	62.41	71	62.29	94	62.06	117	61.75
49	62.41	72	62.28	95	62.05	118	61.74
50	62.40	73	62.27	96	62.04	119	61.72
51	62.40	74	62.26	97	62.02	120	61.71
52	62.40	75	62.25	98	62.01	121	61.69
53	62.39	76	62.25	99	62.00	122	61.68
54	62.39	77	62.24	100	61.99	123	61.66

(continued)

(continued)

Temp., °F	lb per ft³	Temp., °F	lb per ft³	Temp., °F	lb per ft³	Temp., °F	lb per ft³
124	61.64	160	60.99	196	60.21	380	54.47
125	61.63	161	60.97	197	60.19	390	54.05
126	61.61	162	60.95	198	60.16	400	53.62
127	61.60	163	60.93	199	60.14	410	53.19
128	61.58	164	60.91	200	60.11	420	52.74
129	61.56	165	60.89	201	60.09	430	52.33
130	61.55	166	60.87	202	60.07	440	51.87
131	61.53	167	60.85	203	60.04	450	51.28
132	61.51	168	60.83	204	60.02	460	51.02
133	61.50	169	60.81	205	59.99	470	50.51
134	61.48	170	60.79	206	59.97	480	50.00
135	61.46	171	60.77	207	59.95	490	49.50
136	61.44	172	60.75	208	59.92	500	48.78
137	61.43	173	60.73	209	59.90	510	48.31
138	61.41	174	60.71	210	59.87	520	47.62
139	61.39	175	60.68	211	59.85	530	46.95
140	61.37	176	60.66	212	59.82	540	46.30
141	61.36	177	60.64	214	59.81	550	45.66
142	61.34	178	60.62	216	59.77	560	44.84
143	61.32	179	60.60	218	59.70	570	44.05
144	61.30	180	60.57	220	59.67	580	43.29
145	61.28	181	60.55	230	59.42	590	42.37
146	61.26	182	60.53	240	59.17	600	41.49
147	61.25	183	60.51	250	58.89	610	40.49
148	61.23	184	60.49	260	58.62	620	39.37
149	61.21	185	60.46	270	58.34	630	38.31
150	61.19	186	60.44	280	58.04	640	37.17
151	61.17	187	60.42	290	57.74	650	35.97
152	61.15	188	60.40	300	57.41	660	34.48
153	61.13	189	60.37	310	57.08	670	32.89
154	61.11	190	60.35	320	56.75	680	31.06
155	61.09	191	60.33	330	56.40	690	28.82
156	61.07	192	60.30	340	56.02	700	25.38
157	61.05	193	60.28	350	55.65		
158	61.03	194	60.26	360	55.25		
159	61.01	195	60.23	370	54.85		

Feet Head of Water to Pounds per Square Inch*

Feet Head	Pounds per Square Inch	Feet Head	Pounds per Square Inch
1	.43	100	43.31
2	.87	110	47.64
3	1.30	120	51.97
4	1.73	130	56.30
5	2.17	140	60.63
6	2.60	150	64.96
7	3.03	160	69.29
8	3.46	170	76.63
9	3.90	180	77.96
10	4.33	200	86.62
15	6.50	250	108.27
20	8.66	300	129.93
25	10.83	350	151.58
30	12.99	400	173.24
40	17.32	500	216.55
50	21.65	600	259.85
60	25.99	700	303.16
70	30.32	800	346.47
80	34.65	900	389.78
90	38.98	1000	433.00

**One foot of water at 62°F equals 0.433 pound pressure per square inch. To find the pressure per square inch for any feet head not given in the table above, multiply the feet head by 0.433.*

Water Pressure to Feet Head*

Feet Head	Pounds per Square Inch	Feet Head	Pounds per Square Inch
1	2.31	100	230.90
2	4.62	110	253.98
3	6.93	120	277.07
4	9.24	130	300.16
5	11.54	140	323.25
6	13.85	150	346.34
7	16.16	160	369.43
8	18.47	170	392.52
9	20.78	180	415.61
10	23.09	200	461.78
15	34.63	250	577.24
20	46.18	300	692.69
25	57.72	350	808.13
30	69.27	400	922.58
40	92.36	500	1154.48
50	115.45	600	1385.39
60	138.54	700	1616.30
70	161.63	800	1847.20
80	184.72	900	2078.10
90	207.81	1000	2309.00

**One pound of pressure per square inch of water equals 2.309 feet of water at 62°F. Therefore, to find the feet head of water for any pressure not given in the table above, multiply the pressure pounds per square inch by 2.309.*

Boiling Points of Water at Various Pressures

Vacuum, in. Hg	*Boiling Point*	*Vacuum, in. Hg*	*Boiling Point*
29	76.62	7	198.87
28	99.93	6	200.96
27	114.22	5	202.25
26	124.77	4	204.85
25	133.22	3	206.70
24	140.31	2	208.50
23	146.45	1	210.25
22	151.87	Gauge Pounds	
21	156.75	0	212.
20	161.19	1	215.6
19	165.24	2	218.5
18	169.00	4	224.4
17	172.51	6	229.8
16	175.80	8	234.8
15	178.91	10	239.4
14	181.82	15	249.8
13	184.61	25	266.8
12	187.21	50	297.1
11	189.75	75	320.1
10	192.19	100	337.9
9	194.50	125	352.9
8	196.73	200	387.9

Properties of Saturated Steam

Gauge Pressure, psig	Temperature, °F	Heat, Btu/lb			Specific Volume, ft³/lb	Gauge Pressure, psig	Temperature, °F	Heat, Btu/lb			Specific Volume, ft³ per lb
		Sensible	Latent	Total				Sensible	Latent	Total	
25	134	102	1017	1119	142	150	366	339	857	1196	2.74
20	162	129	1001	1130	73.9	155	368	341	885	1196	2.68
15	179	147	990	1137	51.3	160	371	344	853	1197	2.6
10	192	160	982	1142	39.4	165	373	346	851	1197	2.54
5	203	171	976	1147	31.8	170	375	348	849	1197	2.47
0	**212**	**180**	**970**	**1150**	**26.8**	**175**	**377**	**351**	**847**	**1198**	**2.41**
1	215	183	968	1151	25.2	180	380	353	845	1198	2.34
2	219	187	966	1153	23.5	185	382	355	843	1198	2.29
3	222	190	964	1154	22.3	190	384	358	841	1199	2.24
4	224	192	962	1154	21.4	195	386	360	839	1199	2.19
5	**227**	**195**	**960**	**1155**	**20.1**	**200**	**388**	**362**	**837**	**1199**	**2.14**
6	230	198	959	1157	19.4	205	390	364	836	1200	2.09
7	232	200	957	1157	18.7	210	392	366	834	1200	2.05
8	233	201	956	1157	18.4	215	394	368	832	1200	2
9	237	205	954	1159	17.1	220	396	370	830	1200	1.96
10	**239**	**207**	**953**	**1160**	**16.5**	**225**	**397**	**372**	**828**	**1200**	**1.92**
12	244	212	949	1161	15.3	230	399	374	827	1201	1.89
14	248	216	947	1163	14.3	235	401	376	825	1201	1.85

(continued)

(continued)

Gauge Pressure, psig	Temperature, °F	Heat, Btu/lb			Specific Volume, ft³/lb	Gauge Pressure, psig	Temperature, °F	Heat, Btu/lb			Specific Volume, ft³ per lb
		Sensible	Latent	Total				Sensible	Latent	Total	
16	252	220	944	1164	13.4	240	403	378	823	1201	1.81
18	256	224	941	1165	12.6	245	404	380	822	1202	1.78
20	**259**	**227**	**939**	**1166**	**11.9**	**250**	**406**	**382**	**820**	**1202**	**1.75**
22	262	230	937	1167	11.3	255	408	383	819	1202	1.72
24	265	233	934	1167	10.8	260	409	385	817	1202	1.69
26	268	236	933	1169	10.3	265	411	387	815	1202	1.66
28	271	239	930	1169	9.85	270	413	389	814	1203	1.63
30	**274**	**243**	**929**	**1172**	**9.46**	**275**	**414**	**391**	**812**	**1203**	**1.6**
32	277	246	927	1173	9.1	280	416	392	811	1203	1.57
34	279	248	925	1173	8.75	285	417	394	809	1203	1.55
36	282	251	923	1174	8.42	290	418	395	808	1203	1.53
38	284	253	922	1175	8.08	295	420	397	806	1203	1.49
40	**286**	**256**	**920**	**1176**	**7.82**	**300**	**421**	**398**	**805**	**1203**	**1.47**
42	289	258	918	1176	7.57	305	423	400	803	1203	1.45
44	291	260	917	1177	7.31	310	425	402	802	1204	1.43
46	293	262	915	1177	7.14	315	426	404	800	1204	1.41
48	295	264	914	1178	6.94	320	427	405	799	1204	1.38
50	**298**	**267**	**912**	**1179**	**6.68**	**325**	**429**	**407**	**797**	**1204**	**1.36**
55	300	271	909	1180	6.27	330	430	408	796	1204	1.34

(continued)

Gauge Pressure, psig	Temperature, °F	Heat, Btu/lb			Specific Volume, ft^3/lb	Gauge Pressure, psig	Temperature, °F	Heat, Btu/lb			Specific Volume, ft^3 per lb
		Sensible	Latent	Total				Sensible	Latent	Total	
60	307	277	906	1183	5.84	335	432	410	794	1204	1.33
65	312	282	901	1183	5.49	340	433	411	793	1204	1.31
70	316	286	898	1184	5.18	345	434	413	791	1204	1.29
75	320	290	895	1185	4.91	350	435	414	790	1204	1.28
80	324	294	891	1185	4.67	355	437	416	789	1205	1.26
85	328	298	889	1187	4.44	360	438	417	788	1205	1.24
90	331	302	886	1188	4.24	365	440	419	786	1205	1.22
95	335	305	883	1188	4.05	370	441	420	785	1205	1.2
100	338	309	880	1189	3.89	375	442	421	784	1205	1.19
105	341	312	878	1190	3.74	380	443	422	783	1205	1.18
110	344	316	875	1191	3.59	385	445	424	781	1205	1.16
115	347	319	873	1192	3.46	390	446	425	780	1205	1.14
120	350	322	871	1193	3.34	395	447	427	778	1205	1.13
125	353	325	868	1193	3.23	400	448	428	777	1205	1.12
130	356	328	866	1194	3.12	450	460	439	766	1205	1
140	361	333	861	1194	2.92	500	470	453	751	1204	0.89
145	363	336	859	1195	2.84	550	479	464	740	1204	0.82
						600	489	475	728	1203	0.74

Courtesy Sarco Company, Inc.

PIPE PERFORMANCE DATA

Barlow's Formula

Barlow's formula is used to find the relationship between internal fluid pressure and stress in the pipe wall. It is simple to use and it is conservative; the results are safe. Barlow's formula is sometimes known as the *outside diameter* formula because it utilizes the outside diameter of the pipe. Bursting tests on commercial steel pipe of the commonly used thicknesses have shown that Barlow's formula predicts the pressure at which the pipe will rupture with an accuracy well within the limits of uniformity of commercial pipe thickness.

$$P = \frac{2 \times t \times S}{D}$$

Where, P = internal units pressure, psig
S = unit stress, psig
D = outside diameter of pipe, in.
t = wall thickness, in.

Calculating the Linear Expansion of Piping Carrying Steam or Hot Water

Linear Expansion of Piping

Piping carrying steam or hot water will expand or lengthen in direct relation to the temperature of the steam or hot water. The formula for calculating the expansion distance is:

E = constant × (T − F)
E = expansion in inches per hundred feet of pipe
F = starting temperature
T = final temperature

The constants per 100 feet of pipe are given in the following table:

Metal	*Constant*
Steel	0.00804
Wrought Iron	0.00816
Cast Iron	0.00780
Copper-Brass	0.01140

Example:
What is the expansion of 125 feet of steel steam pipe at 10 psig and a starting point of 50°F?

$$E = \text{constant} \times (T - F)$$
$$\text{Constant} = 239.4$$
$$E = 0.00804 \times (239.4 - 50)$$
$$E = 0.00804 \times 189.4$$

```
     0.00804
     × 189.4
    --------
      003216
     007236
    006432
    00804
   --------
   1.522776
```

125 feet of steel steam pipe will expand (lengthen) 1.522 inches at 10 psig with a starting point of 50°F.

Relative Discharging Capacities of Standard Pipe

Pipe Size (in)	Internal Diameter D (in)	Pipe Size in Inches																		
		$D^{5/2}$	⅛	¼	⅜	½	¾	1	1¼	1½	2	2½	3	3½	4	5	6	8	10	12
⅛	0.269	0.037530	1.0	—	—	—	—	—	—	—	—	—	—	—	—	—	—	—	—	—
¼	0.364	0.079938	2.1	1.0	—	—	—	—	—	—	—	—	—	—	—	—	—	—	—	—
⅜	0.493	0.17065	4.5	2.1	1.0	—	—	—	—	—	—	—	—	—	—	—	—	—	—	—
½	0.622	0.30512	8.1	3.8	1.8	1.0	—	—	—	—	—	—	—	—	—	—	—	—	—	—
¾	0.824	0.61634	16	7.7	3.6	2.0	1.0	—	—	—	—	—	—	—	—	—	—	—	—	—
1	1.049	1.1270	30	14	6.6	3.7	3.7	1.0	—	—	—	—	—	—	—	—	—	—	—	—
1¼	1.380	2.2372	60	28	13	7.3	3.6	2.0	1.0	—	—	—	—	—	—	—	—	—	—	—
1½	1.610	3.2890	88	41	19	11	5.3	2.9	1.5	1.0	—	—	—	—	—	—	—	—	—	—
2	2.067	6.1426	164	77	36	20	10	5.5	2.7	1.9	1.0	—	—	—	—	—	—	—	—	—
2½	2.469	9.5786	255	120	56	31	16	8.5	4.3	2.9	1.6	1.0	—	—	—	—	—	—	—	—
3	3.068	16.487	439	206	97	54	27	15	7.4	5.0	2.7	1.7	1.0	—	—	—	—	—	—	—
3½	3.548	23.711	632	297	139	78	38	21	11	7.2	3.9	2.5	1.4	1.0	—	—	—	—	—	—
4	4.026	32.523	867	407	191	107	53	29	15	9.9	5.3	3.4	2.0	1.4	1.0	—	—	—	—	—
5	5.047	57.225	1526	716	335	188	93	51	26	17	9.3	6.0	3.5	2.4	1.8	1.0	—	—	—	—
6	6.065	90.589	2414	1163	531	297	147	80	40	28	15	9.5	5.5	3.8	2.8	1.6	1.0	—	—	—
8	7.981	179.95	4795	2251	1054	590	292	160	80	55	29	19	11	7.6	5.5	3.1	2.0	1.0	—	—
10	10.020	317.81	8468	3976	1862	1042	516	282	142	97	52	33	19	13	9.8	5.6	3.5	1.8	1.0	—
12	12.000	498.83	13292	6240	2923	1635	809	443	223	152	81	52	30	21	15	8.7	5.5	2.8	1.6	1.0

The figure that lies at the intersection of any two sizes is the number of smaller-size pipes required to equal one of the larger.
Example: How many 2-inch standard pipes will it take to equal the discharge of one 8-inch standard pipe?
Solution: Twenty-nine 2-inch pipes; the figure in the table that lies at the intersection of these two sizes is 29.

Pressure Drop in Schedule 40 Pipe

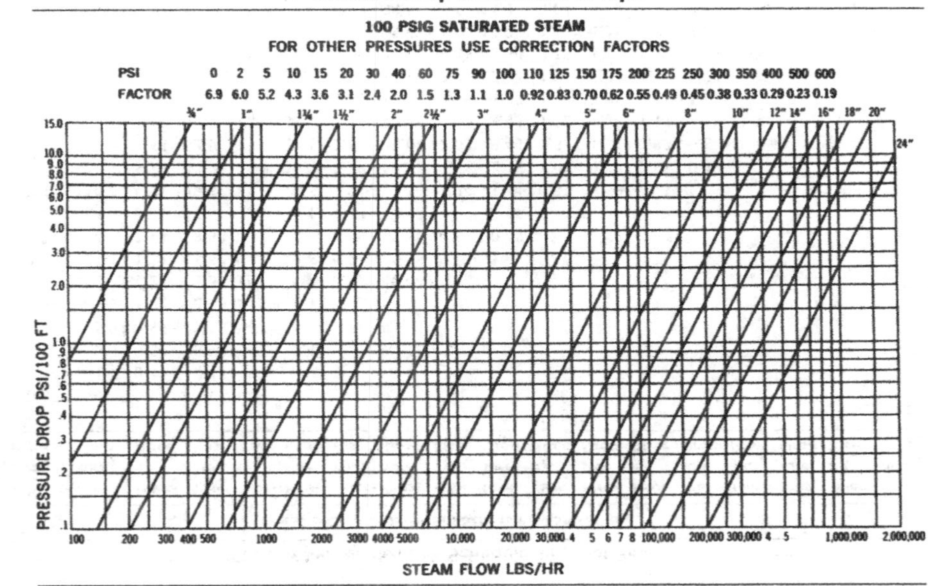

Flow of Water through Schedule 40 Steel Pipe

	Pressure Drop 1000 Feet of Schedule 40 Steel Pipe																	
Dis-charge gal/min	Veloc-ity, ft/sec	Pres-sure Drop	Veloc-ity, ft/sec	Pres-sure Drop	Veloc-ity, ft/sec	Pres-sure Drop	Veloc-ity, ft/sec	Pres-sure Drop	Veloc-ity, ft/sec	Pres-sure Drop	Veloc-ity, ft/sec	Pres-sure Drop	Veloc-ity, ft/sec	Pres-sure Drop	Veloc-ity, ft/sec	Pres-sure Drop	Veloc-ity, ft/sec	Pres-sure Drop
	1"																	
1	0.37	0.49	1¼"															
2	0.74	1.7	0.43	0.45	1½"													
3	1.12	3.53	0.64	0.94	0.47	0.44												
4	1.49	5.94	0.86	1.55	0.63	0.74	2"											
5	1.86	9.02	1.07	2.36	0.79	1.12												
6	2.24	12.25	1.28	3.3	0.95	1.53	0.57	0.46										
8	2.98	21.1	1.72	5.52	1.26	2.63	0.76	0.75	2½"									
10	3.72	30.8	2.14	8.34	1.57	3.86	0.96	1.14	0.67	0.48								
15	5.6	64.6	3.21	17.6	2.36	8.13	1.43	2.33	1	0.99	3"							
20	7.44	110.5	4.29	29.1	3.15	13.5	1.91	3.86	1.34	1.64	0.87	0.59	3½"					
25			5.36	43.7	3.94	20.2	2.39	5.81	1.68	2.48	1.08	0.67	0.81	0.42				
30			6.43	62.9	4.72	29.1	2.87	8.04	2.01	3.43	1.3	1.21	0.97	0.6	4"			
35			7.51	82.5	5.51	38.2	3.35	10.95	2.35	4.49	1.52	1.58	1.14	0.79	0.88	0.42		
40					6.3	47.8	3.82	13.7	2.68	5.88	1.74	2.06	1.3	1	1.01	0.53		
45					7.08	60.6	4.3	17.4	3	7.14	1.95	2.51	1.46	1.21	1.13	0.67		
50					7.87	74.7	4.78	20.6	3.35	8.82	2.17	3.1	1.62	1.44	1.26	0.8		
60							5.74	29.6	4.02	12.2	2.6	4.29	1.95	2.07	1.51	1.1	5"	
70							6.69	38.6	4.69	15.3	3.04	5.84	2.27	2.71	1.76	1.5	1.12	0.48

(continued)

Pressure Drop 1000 Feet of Schedule 40 Steel Pipe

Discharge gal/min	Velocity, ft/sec	Pressure Drop	Velocity, ft/sec	Pressure Drop	Velocity, ft/sec	Pressure Drop	Velocity, ft/sec	Pressure Drop	Velocity, ft/sec	Pressure Drop	Velocity, ft/sec	Pressure Drop	Velocity, ft/sec	Pressure Drop	Velocity, ft/sec	Pressure Drop	Velocity, ft/sec	Pressure Drop
80	6"						7.65	50.3	5.37	21.7	3.48	7.62	2.59	3.53	2.01	1.87	1.28	0.63
90							8.6	63.6	6.04	26.1	3.91	9.22	2.92	4.46	2.26	2.37	1.44	0.8
100	1.11	0.39					9.56	75.1	6.71	32.3	4.34	11.4	3.24	5.27	2.52	2.81	1.6	0.95
125	1.39	0.56							8.38	48.2	5.45	17.1	4.05	7.86	3.15	4.38	2	1.48
150	1.67	0.78							10.06	60.4	6.51	23.3	4.86	11.3	3.78	6.02	2.41	2.04
175	1.94	1.06	8"						11.73	90	7.59	32	5.67	14.7	4.41	8.2	2.81	2.78
200	2.22	1.32									8.68	39.7	6.48	19.2	5.04	10.2	3.21	3.46
225	2.5	1.66	1.44	0.44							9.77	50.2	7.29	23.1	5.67	12.9	3.61	4.37
250	2.78	2.05	1.6	0.55							10.85	61.9	8.1	28.5	6.3	15.9	4.01	5.14
275	3.06	2.36	1.76	0.63							11.94	75	8.91	34.4	6.93	18.3	4.41	6.22
300	3.33	2.8	1.92	0.75							13.02	84.7	9.72	40.9	7.56	21.8	4.81	7.41
325	3.61	3.29	2.08	0.88									10.53	45.5	8.18	25.5	5.21	8.25
350	3.89	3.62	2.24	0.97									11.35	52.7	8.82	29.7	5.61	9.57
375	4.16	4.16	2.4	1.11									12.17	60.7	9.45	32.3	6.01	11
400	4.44	4.72	2.56	1.27									13.78	77.8	10.7	41.5	6.82	14.1
425	4.72	5.34	2.27	1.43									12.97	68.9	10.08	39.7	6.41	12.9
450	5	5.96	2.88	1.6	10"								14.59	87.3	11.33	46.5	7.22	15
475	5.27	6.66	3.04	1.69	1.93	0.3									11.96	51.7	7.62	16.7

(continued)

(continued)

Pressure Drop 1000 Feet of Schedule 40 Steel Pipe																
Dis-charge gal/min	Veloc-ity, ft/sec	Pres-sure Drop	Veloc-ity, ft/sec	Pres-sure Drop	Veloc-ity, ft/sec	Pres-sure Drop	Veloc-ity, ft/sec	Pres-sure Drop	Veloc-ity, ft/sec	Pres-sure Drop	Veloc-ity, ft/sec	Pres-sure Drop	Veloc-ity, ft/sec	Pres-sure Drop	Veloc-ity, ft/sec	Pres-sure Drop
500	5.55	7.39	3.2	1.87	2.04	0.63									12.59	57.3	8.02	18.5
550	6.11	8.94	3.53	2.26	2.24	0.7									13.84	69.3	8.82	22.4
600	6.66	10.6	3.85	2.7	2.44	0.8612"									15.1	82.5	9.62	26.7
650	7.21	11.8	4.17	3.16	2.65	1.01											10.42	31.3
700	7.77	13.7	4.49	3.69	2.85	1.18	2.01	0.48									11.22	36.3
750	8.32	15.7	4.81	4.21	3.05	1.35	2.15	0.55									12.02	41.6
800	8.88	17.8	5.13	4.79	3.26	1.54	2.29	0.62	14"								12.82	44.7
850	9.44	20.2	5.45	5.11	3.46	1.74	2.44	0.7	2.02	0.43							13.62	50.5
900	10	22.6	5.77	5.73	3.66	1.94	2.58	0.79	2.14	0.48							14.42	56.6
950	10.55	23.7	6.09	6.38	3.87	2.23	2.72	0.88	2.25	0.53							15.22	63.1
1,000	11.1	26.3	6.41	7.08	4.07	2.4	2.87	0.98	2.38	0.59							16.02	70
1,100	12.22	31.8	7.05	8.56	4.48	2.74	3.16	1.18	2.61	0.68	16"						17.63	84.6
1,200	13.32	37.8	7.69	10.2	4.88	3.27	3.45	1.4	2.85	0.81	2.18	0.4						
1,300	14.43	44.4	8.33	11.3	5.29	3.86	3.73	1.56	3.09	0.95	2.36	0.47						
1,400	15.54	51.5	8.97	13	5.7	4.44	4.02	1.8	3.32	1.1	2.54	0.54						
1,500	16.65	55.5	9.62	15	6.1	5.11	4.3	2.07	3.55	1.19	2.73	0.62						
1,600	17.76	68.1	10.26	17	6.51	5.46	4.59	2.36	3.8	1.35	2.91	0.71	18"					
1,800	19.98	79.8	11.54	21.6	7.32	6.91	5.16	2.98	4.27	1.71	3.27	0.85	2.58	0.48				

(continued)

Pressure Drop 1000 Feet of Schedule 40 Steel Pipe

Dis-charge gal/min	Veloc-ity, ft/sec	Pres-sure Drop	Veloc-ity, ft/sec	Pres-sure Drop	Veloc-ity, ft/sec	Pres-sure Drop	Veloc-ity, ft/sec	Pres-sure Drop	Veloc-ity, ft/sec	Pres-sure Drop	Veloc-ity, ft/sec	Pres-sure Drop	Veloc-ity, ft/sec	Pres-sure Drop	Veloc-ity, ft/sec	Pres-sure Drop	Veloc-ity, ft/sec	Pres-sure Drop
2,000	22.2	98.5	12.83	25	8.13	8.54	5.73	3.47	4.74	2.11	3.63	1.05	2.88	0.56				
2,500			16.03	39	10.18	12.5	7.17	5.41	5.92	3.09	4.54	1.63	3.59	0.88	20"			
3,000			19.24	52.4	12.21	18	8.6	7.31	7.12	4.45	5.45	2.21	4.31	1.27	3.45	0.73		
3,500			22.43	71.4	14.25	22.9	10.03	9.95	8.32	6.18	6.35	3	5.03	1.52	4.03	0.94	24"	
4,000			25.65	93.3	16.28	29.9	11.48	13	9.49	7.92	7.25	3.92	5.74	2.12	4.61	1.22	3.19	0.51
4,500					18.31	37.8	12.9	15.4	10.67	9.36	8.17	4.97	6.47	2.5	5.19	1.55	3.59	0.6
5,000					20.35	46.7	14.34	18.9	11.84	11.6	9.08	5.72	7.17	3.08	5.76	1.78	3.99	0.74
6,000					24.42	67.2	17.21	27.3	14.32	15.4	10.88	8.24	8.62	4.45	6.92	2.57	4.8	1
7,000					28.5	85.1	20.08	37.2	16.6	21	12.69	12.2	10.04	6.06	8.06	3.5	5.68	1.36
8,000							22.95	45.1	18.98	27.4	14.52	13.6	11.48	7.34	9.23	4.57	6.38	1.78
9,000							25.8	57	21.35	34.7	16.32	17.2	12.92	9.2	10.37	5.36	7.19	2.25
10,000							28.63	70.4	23.75	42.9	18.16	21.2	14.37	11.5	11.53	6.63	7.96	2.78
12,000							34.38	93.6	28.5	61.8	21.8	30.9	17.23	16.5	13.83	9.54	9.57	3.71
14,000									33.2	84	25.42	41.6	20.1	20.7	16.14	12	11.18	5.05
16,000											29.05	54.4	22.96	27.1	18.43	15.7	12.77	6.6

Courtesy Sarco Company, Inc.

Friction Loss for Water in Feet per 100 Feet of Schedule 40 Steel Pipe

U.S. gal/min	Velocity, ft/sec	hf Friction	U.S. gal/min	Velocity, ft/sec	hf Friction
	3/8" Pipe			1/2" Pipe	
1.4	2.25	9.03	2	2.11	5.5
1.6	2.68	11.6	2.5	2.64	8.24
1.8	3.02	14.3	3	3.17	11.5
2	3.36	17.3	3.5	3.7	15.3
2.5	4.2	26	4	4.22	19.7
3	5.04	36.6	5	5.28	29.7
3.5	5.88	49	6	6.34	42
4	6.72	63.2	7	7.39	56
5	8.4	96.1	8	8.45	72.1
6	10.08	136	9	9.5	90.1
7	11.8	182	10	10.56	110.6
8	13.4	236	12	12.7	156
9	15.1	297	14	14.8	211
10	16.8	364	16	16.9	270
	3/4" Pipe			1" Pipe	
4	2.41	4.85	6	2.23	3.16
5	3.01	7.27	8	2.97	5.2
6	3.61	10.2	10	3.71	7.9
7	4.21	13.6	12	4.45	11.1
8	4.81	17.3	14	5.2	14.7
9	5.42	21.6	16	5.94	19
10	6.02	26.5	18	6.68	23.7
12	7.22	37.5	20	7.42	28.9
14	8.42	50	22	8.17	34.8
16	9.63	64.8	24	8.91	41
18	10.8	80.9	26	9.65	47.8
20	12	99	28	10.39	55.1
22	13.2	120	30	11.1	62.9
24	14.4	141	35	13	84.4
26	15.6	165	40	14.8	109
28	16.8	189	45	16.7	137
			50	18.6	168

(continued)

(continued)

U.S. gal/min	Velocity, ft/sec	hf Friction	U.S. gal/min	Velocity, ft/sec	hf Friction
	1¼" Pipe			1½" Pipe	
12	2.57	2.85	16	2.52	2.26
14	3	3.77	18	2.84	2.79
16	3.43	4.83	20	3.15	3.38
18	3.86	6	22	3.47	4.05
20	4.29	7.3	24	3.78	4.76
22	4.72	8.72	26	4.1	5.54
24	5.15	10.27	28	4.41	6.34
26	5.58	11.94	30	4.73	7.2
28	6.01	13.7	35	5.51	9.63
30	6.44	15.6	40	6.3	12.41
35	7.51	21.9	45	7.04	15.49
40	8.58	27.1	50	7.88	18.9
45	9.65	33.8	55	8.67	22.7
50	10.7	41.4	60	9.46	26.7
55	11.8	49.7	65	10.24	31.2
60	12.9	58.6	70	11.03	36
65	13.9	68.6	75	11.8	41.2
70	15	79.2	80	12.6	46.6
75	16.1	90.6	85	13.4	52.4
			90	14.2	58.7
			95	15	65
			100	15.8	71.6
	2" Pipe			2½" Pipe	
25	2.39	1.48	35	2.35	1.15
30	2.87	2.1	40	2.68	1.47
35	3.35	2.79	45	3.02	1.84
40	3.82	3.57	50	3.35	2.23
45	4.3	4.4	60	4.02	3.13
50	4.78	5.37	70	4.69	4.18
60	5.74	7.58	80	5.36	5.36
70	6.69	10.2	90	6.03	6.69
80	7.65	13.1	100	6.7	8.18
90	8.6	16.3	120	8.04	11.5

(continued)

(continued)

U.S. gal/min	Velocity, ft/sec	hf Friction	U.S. gal/min	Velocity, ft/sec	hf Friction
	2" Pipe			**2½" Pipe**	
100	4.34	2.72	200	5.04	12.61
120	11.5	28.5	160	10.7	20
140	13.4	38.2	180	12.1	25.2
160	15.3	49.5	200	13.4	30.7
			220	14.7	37.1
			240	16.1	43.8
	3" Pipe			**4" Pipe**	
50	2.17	0.762	100	2.52	0.718
60	2.6	1.06	120	3.02	1.01
70	3.04	1.4	140	3.53	1.35
80	3.47	1.81	160	4.03	1.71
90	3.91	2.26	180	4.54	2.14
100	3.34	2.75	200	5.04	2.61
120	5.21	3.88	220	5.54	3.13
140	6.08	5.19	240	6.05	3.7
160	6.94	6.68	260	6.55	4.3
180	7.81	8.38	280	7.06	4.95
200	8.68	10.2	300	7.56	5.63
220	9.55	12.3	350	8.82	7.54
240	10.4	14.5	400	10.1	9.75
260	11.3	16.9	450	11.4	12.3
280	12.2	19.5	500	12.6	14.4
300	13	22.1	550	13.9	18.1
350	15.2	30	600	15.1	21.4
	5" Pipe			**6" Pipe**	
160	2.57	0.557	220	2.44	0.411
180	2.89	0.698	240	2.66	0.482
200	3.21	0.847	260	2.89	0.56
220	3.53	1.01	300	3.33	0.733
240	3.85	1.19	350	3.89	0.98
260	4.17	1.38	400	4.44	1.25
300	4.81	1.82	450	5	1.56

(continued)

(continued)

U.S. gal/min	Velocity, ft/sec	hf Friction	U.S. gal/min	Vel, ft/sec	hf Friction
350	5.61	2.43	500	5.55	1.91
400	6.41	3.13	600	6.66	2.69
450	7.22	3.92	700	7.77	3.6
500	8.02	4.79	800	8.88	4.64
600	9.62	6.77	900	9.99	5.81
700	11.2	9.13	1000	11.1	7.1
800	12.8	11.8	1100	12.2	8.52
900	14.4	14.8	1200	13.3	10.1
1000	16	18.2	1300	14.4	11.7
			1400	15.5	13.6

Courtesy Sarco Company, Inc.

Pressure Drop in Schedule 80 Pipe

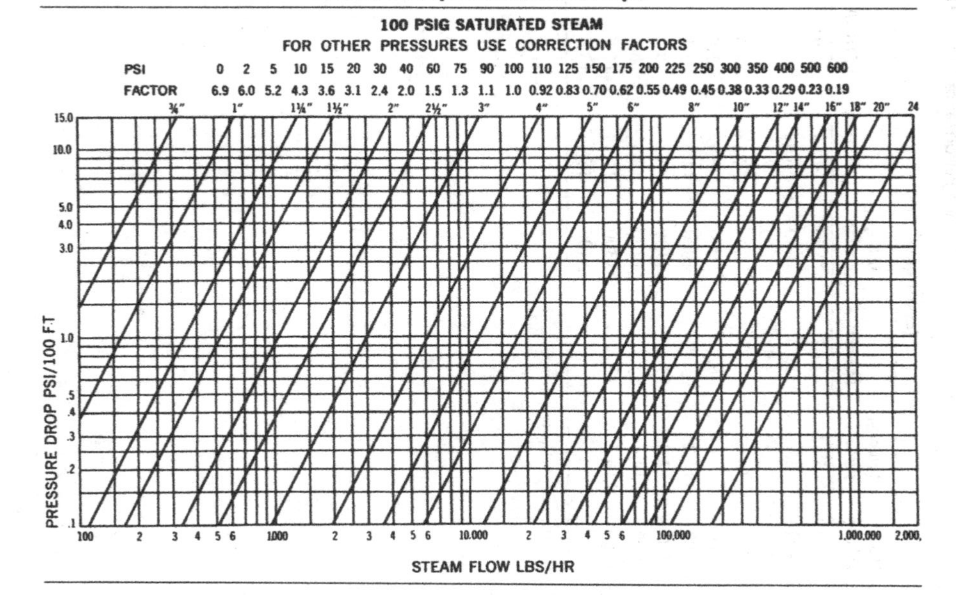

Warmup Load in Pounds of Steam per 100 Feet of Steam Main (ambient temperature 70°F)*†

Steam Pressure	Main Size														0°F Correction
psig	2"	2½"	3"	4"	5"	6"	8"	10"	12"	14"	16"	18"	20"	24"	Factor
0	6.2	9.7	12.8	18.2	24.6	31.9	48	68	90	107	140	176	207	208	1.5
5	6.9	11	14.4	20.4	27.7	35.9	48	77	101	120	157	198	233	324	1.44
10	7.5	11.8	15.5	22	29.9	38.8	58	83	109	130	169	213	251	350	1.41
20	8.4	13.4	17.5	24.9	33.8	43.9	66	93	124	146	191	241	284	396	1.37
40	3.9	15.8	20.6	29.3	39.7	51.6	78	110	145	172	225	284	334	465	1.32
60	11	17.5	22.9	32.6	44.2	57.3	86	122	162	192	250	316	372	518	1.29
80	12	19	24.9	35.3	47.9	62.1	93	132	175	208	271	342	403	561	1.27
100	12.8	20.3	26.6	37.8	51.2	66.5	100	142	188	222	290	366	431	600	1.26
125	13.7	21.7	28.4	40.4	54.8	71.1	107	152	200	238	310	391	461	642	1.25
150	14.5	23	30	42.8	58	75.2	113	160	212	251	328	414	487	679	1.24
175	15.3	24.2	31.7	45.1	61.2	79.4	119	169	224	265	347	437	514	716	1.23
200	16	25.3	33.1	47.1	63.8	82.8	125	177	234	277	362	456	537	748	1.22
250	17.2	27.3	35.8	50.8	68.9	89.4	134	191	252	299	390	492	579	807	1.21
300	25	38.3	51.3	74.8	104	142.7	217	322	443	531	682	854	1045	1182	1.2
400	27.8	42.6	57.1	83.2	115.7	158.7	241	358	493	590	759	971	1163	1650	1.18
500	30.2	46.3	62.1	90.5	125.7	172.6	262	389	535	642	825	1033	1263	1793	1.17
600	32.7	50.1	67.1	97.9	136	186.6	284	421	579	694	893	1118	1367	1939	1.16

**Loads based on Schedule 40 pipe for pressures up to and including 250 psig and on Schedule 80 pipe for pressures above 250 psig.*

†For outdoor temperature of 0°F, multiply load value in table for each main size by correction factor corresponding to steam pressure.

Courtesy Sarco Company, Inc.

Condensation Load in Pounds per Hour per 100 Feet of Insulated Steam Main (ambien temperature 70°F; insulat on 80% efficient)*†

Steam Pressure, psig	Main Size														0°F Correction Factor
	2"	2½"	3"	4"	5"	6"	8"	10"	12"	14"	16"	18"	20"	24"	
10	6	7	9	11	13	16	20	24	29	32	36	39	44	53	1.58
30	8	9	11	14	17	20	26	32	38	42	48	51	57	68	1.5
60	10	12	14	18	24	27	33	41	49	54	62	67	74	89	1.45
100	12	15	18	22	28	33	41	51	61	67	77	83	93	111	1.41
125	13	16	20	24	30	36	45	56	66	73	84	90	101	121	1.39
175	16	19	23	26	33	38	53	66	78	86	98	107	119	142	1.38
250	18	22	27	34	42	50	62	77	92	101	116	126	140	168	1.36
300	20	25	30	37	46	54	68	85	101	111	126	138	154	184	1.35
400	23	28	34	43	53	63	80	99	118	130	148	162	180	216	1.33
500	27	33	39	49	61	73	91	114	135	148	170	185	206	246	1.32
600	30	37	44	55	68	82	103	128	152	167	191	208	232	277	1.31

Chart loads represent losses due to radiation and convection for saturated steam.

†*For outdoor temperature of 0°F, multiply load value in table for each main size by correction factor corresponding to steam pressure.*

Courtesy Sarco Company, Inc.

Expansion of Steam Pipes (inches increase per 100 feet)

Temperature, °F	*Steel*	*Wrought Iron*	*Cast Iron*	*Brass and Copper*
0	0	0	0	0
20	0.15	0.15	0.10	0.25
40	0.30	0.30	0.30	0.45
60	0.45	0.45	0.40	0.65
80	0.60	0.60	0.55	0.90
100	0.75	0.80	0.75	1.15
120	0.90	0.95	0.85	1.40
140	1.10	1.15	1.00	1.65
160	1.25	1.35	1.15	1.90
180	1.45	1.50	1.30	2.15
200	1.60	1.65	1.50	2.40
220	1.80	1.85	1.65	2.65
240	2.00	2.05	1.80	2.90
260	2.15	2.20	1.95	3.15
280	2.35	2.40	2.15	3.45
300	2.50	2.60	2.35	3.75
320	2.70	2.80	2.50	4.05
340	2.90	3.05	2.70	4.35
360	3.05	3.25	2.90	4.65
380	3.25	3.45	3.10	4.95
400	3.45	3.65	3.30	5.25
420	3.70	3.90	3.50	5.60
440	3.95	4.20	3.75	5.95
460	4.20	4.45	4.00	6.30
480	4.45	4.70	4.25	6.65
500	4.70	4.90	4.45	7.05
520	4.95	5.15	4.70	7.45
540	5.20	5.40	4.95	7.85
560	5.45	5.70	5.20	8.25
580	5.70	6.00	5.45	8.65
600	6.00	6.25	5.70	9.05
620	6.30	6.55	5.95	9.50
640	6.55	6.85	6.25	9.95
660	6.90	7.20	6.55	10.40

(continued)

(continued)

Temperature, °F	*Steel*	*Wrought Iron*	*Cast Iron*	*Brass and Copper*
680	7.20	7.50	6.85	10.95
700	7.50	7.85	7.15	11.40
720	7.80	8.20	7.45	11.90
740	8.20	8.55	7.80	12.40
760	8.55	8.90	8.15	12.95
780	8.95	9.30	8.50	13.50
800	9.30	9.75	8.90	14.10

Safe Strength of Soldered Joints—Pressure Ratings, Maximum Service Pressure (psi)

		Water‡		
Solder Used in Joints	*Service Temperatures °F*	*¼ to 1 in. Incl.**	*1¼ to 2 In. Incl.**	*2½ to 4 In. Incl.**
50-50 tin-lead†	100	200	175	150
	150	150	125	100
	200	100	90	75
	250	85	75	50
95-5 tin-antimony	100	500	400	300
	150	400	350	275
	200	300	250	200
	250	200	175	150
Brazing filler metal melting at or above 1000°F§	250	300	210	170
	350	270	190	155

**Standard copper water tube sizes*

†ASTM B32, alloy grade 50A

‡Including refrigerants and other noncorrosive liquids and gases

§ASTM B260, brazing filler metal

Courtesy Copper & Brass Research Assoc.

Wrought-Steel Pipe—Theoretical Bursting and Working Pressures (pounds per square inch)*

		Standard		Extra Strong		Double Extra Strong		Large OD.			
								⅜-Inch-		½-Inch-	
Size, (in)	Size, (mm)	Bursting Pressure Barlow's Formula	Working Pressure Factor 8	Bursting Pressure Barlow's Formula	Working Pressure Factor 8	Bursting Pressure Barlow's Formula	Working Pressure Factor 8	Thick Bursting Pressure Barlow's Formula	Working Pressure Factor 8	Thick Bursting Pressure Barlow's Formula	Working Pressure Factor 8
⅛	3	13,432	1679	18,760	2345						
¼	6	13,032	1629	17,624	2204						
⅜	10	10,784	1348	14,928	1866						
½	13	10,384	1298	14,000	1750	28,000	3500				
¾	19	8,608	1076	11,728	1466	23,464	2933				
1	25	8,088	1011	10,888	1361	21,776	2722				
1¼	32	6,744	843	9,200	1150	18,408	2301				
1½	38	6,104	763	8,416	1052	16,840	2105				
2	50	5,184	648	7,336	917	14,680	1835				
2½	64	5,648	706	7,680	960	15,360	1920				
3	76	4,936	617	6,856	857	13,714	1714				
3½	90	5,610	701	7,950	994	15,900	1987				
4	100	5,266	658	7,480	935	14,970	1871				
4½	113	4,940	618	7,100	887	14,200	1775				
5	125	4,630	579	6,740	842	13,480	1685				
6	150	4,220	528	6,520	815	13,040	1630				

(continued)

		Standard		Extra Strong		Double Extra Strong		Large OD.			
								3/8-Inch-		1/2-Inch-	
Size, (in)	Size, (mm)	Bursting Pressure Barlow's Formula	Working Pressure Factor 8	Bursting Pressure Barlow's Formula	Working Pressure Factor 8	Bursting Pressure Barlow's Formula	Working Pressure Factor 8	Thick Bursting Pressure Barlow's Formula	Working Pressure Factor 8	Thick Bursting Pressure Barlow's Formula	Working Pressure Factor 8
7	175	3,940	493	6,550	819	11,470	1434				
8	200	3,730	466	5,780	722	10,140	1267				
9	225	3,550	444	5,190	649						
10	300	3,390	424	4,650	581						
12	300	2,940	368	3,920	490						
14	350							2,680	335	3,570	446
15	375							2,500	313	3,333	417
16	400							2,340	293	3,120	390
18	450							2,080	260	2,770	346
20	500							1,870	234	2,500	313
22	550							1,700	213	2,270	284
24	600							1,560	195	2,080	260

*Butt-welded pipe was figured on sizes 3 inches and smaller and lap-welded pipe on sizes 3 1/2 inches and larger.

COPPER TUBE

Four different types of copper tube are used in HVAC and domestic water applications: Type K, Type L, Type M, and ACR. Types K, L, and M are designated by ASTM standard sizes, with the outside diameter (OD) always 1/8 inch larger than the standard size designation. Each type represents a series of sizes with different wall thicknesses. Type K copper tube has the thickest walls, Type L has thinner walls than Type K, and Type M has the thinnest walls. The inside diameters depend on tube size and wall thickness.

Seamless copper water tube is manufactured in sizes 1/4 inch through 12 inches nominal. Types K and L are manufactured in drawn temper (hard) tubing in diameters ranging from 1/4 inch through 12 inches and annealed temper (soft) coils with diameters ranging from 1/4 inch through 2 inches. Type M is manufactured only in drawn (hard) temper tube, and is available in diameters ranging from 1/4 inch through 12 inches.

Drawn temper (hard) seamless copper water tube must be identified with a color stripe containing the manufacturer's name or trademark, type of tube, and nation of origin. Type K tube is identified by a green stripe, Type L by a blue stripe, and Type M by a red one. Note that annealed (soft) coils or annealed straight lengths are not required to be identified with a color stripe. Drawn temper copper tube also must be permanently marked (incised) in accordance with governing specifications to show tube type, the name or trademark of the manufacturer, and the country of origin. The incised information must be repeated along the length of the tubing in intervals not to exceed 1½ feet.

The copper tube used in air conditioning and refrigeration systems in the field must meet the requirements of ASTM B 280 (*Standard Specification for Seamless Copper Tube for Air Conditioning and Refrigeration Field Service*). It is sometimes referred to as *refer* or *ACR* tube. Note the following:

- The name or trademark of the tube manufacturer and the ACR mark must be incised on straight lengths of tube at intervals not greater than 1½ feet along the length of the tube. Hard straight lengths of copper tube should also be marked with a blue stripe containing the tube manufacturer's name or trademark, the nation of origin, outside diameter, and the word "ACR" at intervals of 3 feet.
- The name or trademark of the tube manufacturer and the ACR mark must be incised on copper tube coils of ¼-inch diameter or larger at intervals not greater than 1½ feet.
- ASTM B280 copper tube is produced in straight lengths in sizes ranging from ⅜ inch OD through 4⅛ inches OD. It is also produced in annealed tube coils ranging in sizes from ⅛ inch OD through 1⅝ inches OD ACR.
- Copper tube (ACR) used in HVAC/R systems is designated by its actual outside diameter.
- Coiled or straight-length ASTM B 280 tube must be cleaned and capped before shipping.

Both Type L and ACR copper tube can be used for fuel oil, propane (LP), and natural gas service. Flared joints should be used to join tube lengths in accessible locations. Use brazed joints made with AWS A5.8 BAg series brazing filler metals to join Type L or ACR tube in concealed locations.

The word *temper* is used to describe the strength and hardness of a copper tube. Drawn temper copper tube is frequently referred to as *hard* tube. Annealed copper tube is sometimes called *soft* tube. Hard and soft temper tube can be joined by soldering or brazing, using capillary fittings, by welding, or by a variety of different mechanical techniques.

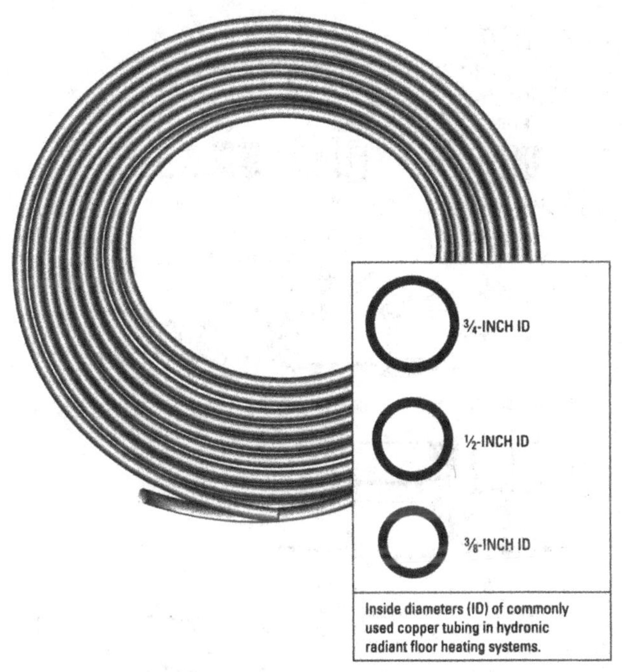

Copper tubing.

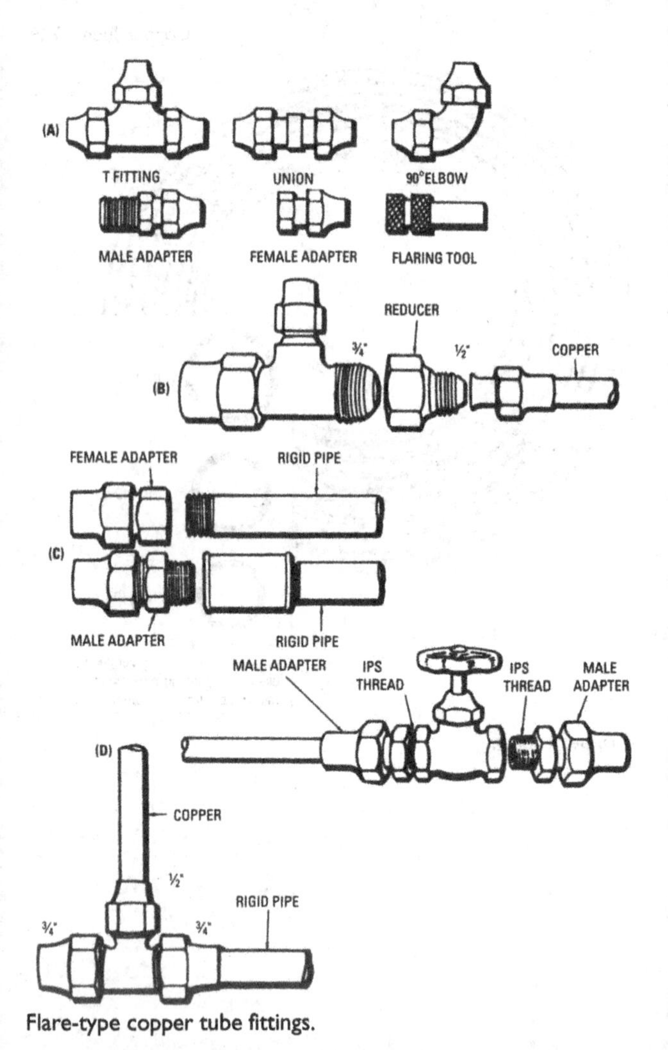

Flare-type copper tube fittings.

Copper Tube: Types, Standards, Applications, Tempers, Lengths

Tube Type	Color Code	Standard	Application*	Commercially Available Lengths[†] Nominal or Standard Sizes	Drawn	Annealed
				STRAIGHT LENGTHS:		
Type K	Green	ASTM B 88[‡]	Domestic water service and Distribution, fire protection, Solar, Fuel/Fuel Oil, HVAC, Snow Melting, Compressed Air, Natural Gas, Liquified Petroleum (LP) Gas, Vacuum	½-inch to 8-inch	20 ft	20 ft
				10-inch	18 ft	18 ft
				12-inch	12 ft	12 ft
				COILS:		
				¼-inch to 1-inch	—	60 ft
					—	100 ft
				1¼ inch and 1½-inch	—	60 ft
				2-inch	—	40 ft
					—	45 ft

(continued)

(continued)

Tube Type	Color Code	Standard	Application*	Commercially Available Lengths† Nominal or Standard Sizes	Drawn	Annealed
				STRAIGHT LENGTHS:		
Type L	Blue	ASTM B 88	Domestic Water Service and Distribution, Fire Protection, Solar, Fuel/Fuel Oil, Natural Gas, Liquified Petroleum (LP) Gas, HVAC, Snow Melting, Compressed Air, Vacuum	½-inch to 10-inch	20 ft	20 ft
				12-inch	18 ft	18 ft
				COILS:		
				¼-inch to 1-inch	—	60 ft
					—	100 ft
				1¼ inch and 1½-inch	—	60 ft
				2-inch	—	40 ft
					—	45 ft

				STRAIGHT LENGTHS:		
Type M	Red	ASTM B 88	Domestic Water Service and Distribution, Fire Protection, Solar, Fuel/Fuel Oil, HVAC, Snow Melting, Vacuum	¼-inch to 12-inch	20 ft	N/A
				STRAIGHT LENGTHS:		
DWV	Yellow	ASTM B 306	Drain, Waste, Vent, HVAC, Solar	1¼-inch to 8-inch	20 ft	N/A
				STRAIGHT LENGTHS:		
ACR	Blue	ASTM B 280	Air Conditioning, Refrigeration, Natural Gas, Liquified Petroleum (LP) Gas, Compressed Air	⅜-inch to 4⅛-inch	20 ft	§
				COILS:		
				⅛-inch to 1⅝-inch	—	50 ft

(continued)

(continued)

Tube Type	Color Code	Standard	Application*	Commercially Available Lengths† Nominal or Standard Sizes	Drawn	Annealed
OXY, MED,	(K) Green			STRAIGHT LENGTHS:		
OXY/MED,		ASTM B 819	Medical Gas	¼-inch to 8-inch	20 ft	N/A
OXY/ACR,	(L) Blue		Compressed			
ACR/MED			Medical Air,			
			Vacuum			

*There are many other copper and copper-alloy tubes and pipes available for specialized applications. For information on these products, contact the Copper Development Association, Inc.

†Individual manufacturers may have commercially available lengths in addition to those shown in this table.

‡Tube made to other ASTM standards is also intended for plumbing applications, although ASTM B 88 is by far the most widely used. ASTM Standard Classification B 698 lists six plumbing tube standards, including B 88.

§Available as special order only.

Courtesy Copper Development Association

Dimensions and Physical Characteristics of Type K Copper Tube

Nominal or Standard Size, in.	Nominal Dimensions, in.			Calculated Values (based on nominal dimensions)				
	Outside Diameter	Inside Diameter	Wall Thickness	Cross Sectional Area of Bore, in.^{2}	Weight of Tube Only, lbs per linear ft	Weight of Tube & Water, lbs per linear ft	Contents of Tube per linear ft, ft^3	Contents of Tube per linear ft, gal
¼	0.375	0.305	0.035	0.073	0.145	0.177	0.00051	0.00379
⅜	0.500	0.402	0.049	0.127	0.269	0.324	0.00088	0.00660
½	0.625	0.527	0.049	0.218	0.344	0.438	0.00151	0.0113
⅝	0.750	0.652	0.049	0.334	0.418	0.562	0.00232	0.0174
¾	0.875	0.745	0.065	0.436	0.641	0.829	0.00303	0.0227
1	1.125	0.992	0.065	0.778	0.839	1.18	0.00540	0.0404
1¼	1.375	1.245	0.065	1.22	1.04	1.57	0.00847	0.0634
1½	1.625	1.481	0.072	1.72	1.36	2.10	0.0119	0.0894
2	2.125	1.959	0.083	3.01	2.06	3.36	0.0209	0.156
2½	2.625	2.435	0.095	4.66	2.93	4.94	0.0324	0.242
3	3.125	2.907	0.109	6.64	4.00	6.87	0.0461	0.345
3½	3.625	3.385	0.120	9.00	5.12	9.01	0.0625	0.468
4	4.125	3.857	0.134	11.7	6.51	11.6	0.0813	0.608

(continued)

(continued)

Nominal or Standard Size, in.	Nominal Dimensions, in.			Calculated Values (based on nominal dimensions)				
	Outside Diameter	Inside Diameter	Wall Thickness	Cross Sectional Area of Bore, $in.^2$	Weight of Tube Only, lbs per linear ft	Weight of Tube & Water, lbs per linear ft	Contents of Tube per linear ft ft^3	Contents of Tube per linear ft gal
5	5.125	4.805	0.160	18.1	9.67	17.5	0.126	0.940
6	6.125	5.741	0.192	25.9	13.9	25.1	0.180	1.35
8	8.125	7.583	0.271	45.2	25.9	45.4	0.314	2.35
10	10.125	9.449	0.338	70.1	40.3	70.6	0.487	3.64
12	12.125	11.315	0.405	101	57.8	101	0.701	5.25

Courtesy Copper Development Association

Dimensions and Physical Characteristics of Type L Copper Tube

Nominal or Standard Size, in.	Nominal Dimensions, in.			Calculated Values (based on nominal dimensions)				
	Outside Diameter	Inside Diameter	Wall Thickness	Cross Sectional Area of Bore, $in.^2$	Weight of Tube Only, lbs per linear ft	Weight of Tube & Water, lbs per linear ft	Contents of Tube per linear ft, ft^3	Contents of Tube per linear ft, gal
¼	0.375	0.315	0.030	0.078	0.126	0.160	0.00054	0.00405
⅜	0.500	0.430	0.035	0.145	0.198	0.261	0.00101	0.00753
½	0.625	0.545	0.040	0.233	0.285	0.386	0.00162	0.0121
⅝	0.750	0.666	0.042	0.348	0.362	0.506	0.00232	0.0174
¾	0.875	0.785	0.045	0.484	0.455	0.664	0.00336	0.0251
1	1.125	1.025	0.050	0.825	0.655	1.01	0.00573	0.0429
1¼	1.375	1.265	0.055	1.26	0.884	1.43	0.00875	0.0655
1½	1.625	1.505	0.060	1.78	1.14	1.91	0.0124	0.0925
2	2.125	1.985	0.070	3.09	1.75	3.09	0.0215	0.161
2½	2.625	2.465	0.080	4.77	2.48	4.54	0.0331	0.248
3	3.125	2.945	0.090	6.81	3.33	6.27	0.0473	0.354
3½	3.625	3.425	0.100	9.21	4.29	8.27	0.0640	0.478
4	4.125	3.905	0.110	12.0	5.38	10.1	0.0764	0.571

(continued)

(continued)

Nominal or Standard Size, in.	Nominal Dimensions, in.			Calculated Values (based on nominal dimensions)				
	Outside Diameter	Inside Diameter	Wall Thickness	Cross Sectional Area of Bore, $in.^2$	Weight of Tube Only, lbs per linear ft	Weight of Tube & Water, lbs per linear ft	Contents of Tube per linear ft	
							ft^3	gal
5	5.125	4.875	0.125	18.7	7.61	15.7	0.130	0.971
6	6.125	5.845	0.140	26.8	10.2	21.8	0.186	1.39
8	8.125	7.725	0.200	46.9	19.3	39.6	0.326	2.44
10	10.125	9.625	0.250	72.8	30.1	61.6	0.506	3.78
12	12.125	11.565	0.280	105	40.4	85.8	0.729	5.45

Courtesy Copper Development Association

Dimensions and Physical Characteristics of Type M Copper Tube

Nominal or Standard Size, in.	Nominal Dimensions, in.			Calculated Values (based on nominal dimensions)				
	Outside Diameter	Inside Diameter	Wall Thickness	Cross Sectional Area of Bore, in.^{2}	Weight of Tube Only, lbs per linear ft	Weight of Tube & Water, lbs per linear ft	Contents of Tube per linear ft: ft^3	Contents of Tube per linear ft: gal
3/8	0.500	0.450	0.025	0.159	0.145	0.214	0.00110	0.00826
1/2	0.625	0.569	0.028	0.254	0.204	0.314	0.00176	0.0132
3/4	0.875	0.811	0.032	0.517	0.328	0.551	0.00359	0.0269
1	1.125	1.055	0.035	0.874	0.465	0.843	0.00607	0.0454
1 1/4	1.375	1.291	0.042	1.31	0.682	1.25	0.00910	0.0681
1 1/2	1.625	1.527	0.049	1.83	0.940	1.73	0.0127	0.0951
2	2.125	2.009	0.058	3.17	1.46	2.83	0.0220	0.165
2 1/2	2.625	2.495	0.065	4.89	2.03	4.14	0.0340	0.254
3	3.125	2.981	0.072	6.98	2.68	5.70	0.0485	0.363
3 1/2	3.625	3.459	0.083	9.40	3.58	7.64	0.0653	0.488
4	4.125	3.935	0.095	12.2	4.66	9.83	0.0847	0.634
5	5.125	4.907	0.109	18.9	6.66	14.8	0.131	0.982
6	6.125	5.881	0.122	27.2	8.92	20.7	0.189	1.41

(continued)

(continued)

Nominal or Standard Size, in.	Nominal Dimensions, in.			Calculated Values (based on nominal dimensions)				
	Outside Diameter	Inside Diameter	Wall Thickness	Cross Sectional Area of Bore, $in.^2$	Weight of Tube Only, lbs per linear ft	Weight of Tube & Water, lbs per linear ft	Contents of Tube per linear ft	
							ft^3	gal
8	8.125	7.785	0.170	47.6	16.5	37.1	0.331	2.47
10	10.125	9.701	0.212	73.9	25.6	57.5	0.513	3.84
12	12.125	11.617	0.254	106	36.7	82.5	0.736	5.51

Courtesy Copper Development Association

Dimensions and Physical Characteristics of Copper Tube: ACR (air conditioning and refrigeration field service) (A = annealed temper, D = drawn temper)

Nominal or Standard Size, in.		Nominal Dimensions, in. Outside Diameter	Inside Diameter	Wall Thickness	Calculated Values (based on nominal dimensions) Cross Sectional Area of Bore, in.^{2}	External Surface ft^2 per linear ft	Internal Surface, ft^2 per linear ft	Weight of Tube Only lbs per linear ft	Contents of Tube ft^2 per linear ft
1/8	A	0.125	0.065	0.030	0.00332	0.0327	0.0170	0.0347	0.00002
3/16	A	0.187	0.128	0.030	0.0129	0.0492	0.0335	0.0575	0.00009
1/4	A	0.250	0.190	0.030	0.0284	0.0655	0.0497	0.0804	0.00020
5/16	A	0.312	0.248	0.032	0.0483	0.0817	0.0649	0.109	0.00034
3/8	A	0.375	0.311	0.032	0.076	0.0982	0.0814	0.134	0.00053
	D	0.375	0.315	0.030	0.078	0.0982	0.0821	0.126	0.00054
1/2	A	0.500	0.436	0.032	0.149	0.131	0.114	0.182	0.00103
	D	0.500	0.430	0.035	0.145	0.131	0.113	0.198	0.00101
5/8	A	0.625	0.555	0.035	0.242	0.164	0.145	0.251	0.00168
	D	0.625	0.545	0.040	0.233	0.164	0.143	0.285	0.00162
3/4	A	0.750	0.680	0.035	0.363	0.196	0.178	0.305	0.00252
	A	0.750	0.666	0.042	0.348	0.196	0.174	0.362	0.00242
	D	0.750	0.666	0.042	0.348	0.196	0.174	0.362	0.00242

(continued)

(continued)

Nominal or Standard Size, in.		Nominal Dimensions, in.			Calculated Values (based on nominal dimensions)				
		Outside Diameter	Inside Diameter	Wall Thickness	Cross Sectional Area of Bore, in.^{2}	External Surface ft^2 per linear ft	Internal Surface, ft^2 per linear ft	Weight of Tube Only lbs per linear ft	Contents of Tube ft^2 per linear ft
7/8	A	0.875	0.785	0.045	0.484	0.229	0.206	0.455	0.00336
	D	0.875	0.785	0.045	0.484	0.229	0.206	0.455	0.00336
1 1/8	A	1.125	1.025	0.050	0.825	0.294	0.268	0.655	0.00573
	D	1.125	1.025	0.050	0.825	0.294	0.268	0.655	0.00573
1 3/8	A	1.375	1.265	0.055	1.26	0.360	0.331	0.884	0.00875
	D	1.375	1.265	0.055	1.26	0.360	0.331	0.884	0.00875
1 5/8	A	1.625	1.505	0.060	1.78	0.425	0.394	1.14	0.0124
	D	1.625	1.505	0.060	1.78	0.425	0.394	1.14	0.0124
2 1/8	D	2.125	1.985	0.070	3.09	0.556	0.520	1.75	0.0215
2 5/8	D	2.625	2.465	0.080	4.77	0.687	0.645	2.48	0.0331
3 1/8	D	3.125	2.945	0.090	6.81	0.818	0.771	3.33	0.0473
3 5/8	D	3.625	3.425	0.100	9.21	0.949	0.897	4.29	0.0640
4 1/8	D	4.125	3.905	0.110	12.0	1.08	1.02	5.38	0.0833

Courtesy Copper Development Association

Rated Internal Working Pressure for Type K Copper Tube

Nominal Size, in.	Annealed							Drawn*						
	S = 6000 psi 100°F	S = 5100 psi 150°F	S = 4900 psi 200°F	S = 4800 psi 250°F	S = 4700 psi 300°F	S = 4000 psi 350°F	S = 3000 psi 400°F	S = 10,300 psi 100°F	S = 10,300 psi 150°F	S = 10,300 psi 200°F	S = 10,300 psi 250°F	S = 10,000 psi 300°F	S = 9700 psi 350°F	S = 9400 psi 400°F
¼	1074	913	877	860	842	716	537	1850	1850	1850	1850	1796	1742	1688
⅜	1130	960	923	904	885	753	565	1946	1946	1946	1946	1889	1833	1776
½	891	758	728	713	698	594	446	1534	1534	1534	1534	1490	1445	1400
⅝	736	626	601	589	577	491	368	1266	1266	1266	1266	1229	1193	1156
¾	852	724	696	682	668	568	426	1466	1466	1466	1466	1424	1381	1338
1	655	557	535	524	513	437	327	1126	1126	1126	1126	1093	1061	1028
1¼	532	452	434	425	416	354	266	914	914	914	914	888	861	834
1½	494	420	404	396	387	330	247	850	850	850	850	826	801	776
2	435	370	355	348	341	290	217	747	747	747	747	726	704	682
2½	398	338	325	319	312	265	199	684	684	684	684	664	644	624
3	385	328	315	308	302	257	193	662	662	662	662	643	624	604
3½	366	311	299	293	286	244	183	628	628	628	628	610	592	573
4	360	306	294	288	282	240	180	618	618	618	618	600	582	564

(continued)

(continued)

Nominal Size, in.	Annealed							Drawn*						
	S = 6000 psi 100°F	S = 5100 psi 150°F	S = 4900 psi 200°F	S = 4800 psi 250°F	S = 4700 psi 300°F	S = 4000 psi 350°F	S = 3000 psi 400°F	S = 10,300 psi 100°F	S = 10,300 psi 150°F	S = 10,300 psi 200°F	S = 10,300 psi 250°F	S = 10,000 psi 300°F	S = 9700 psi 350°F	S = 9400 psi 400°F
5	345	293	281	276	270	230	172	592	592	592	592	575	557	540
6	346	295	283	277	271	231	173	595	595	595	595	578	560	543
8	369	314	301	295	289	246	184	634	634	634	634	615	597	578
10	369	314	301	295	289	246	184	634	634	634	634	615	597	578
12	370	314	302	296	290	247	185	635	635	635	635	617	598	580

When brazing or welding is used to join drawn tube, the corresponding annealed rating must be used.

Courtesy Copper Development Association

Rated Internal Working Pressure for Type L Copper Tube

	Annealed							Drawn*						
Nominal Size, in.	S = 6000 psi 100°F	S = 5100 psi 150°F	S = 4900 psi 200°F	S = 4800 psi 250°F	S = 4700 psi 300°F	S = 4000 psi 350°F	S = 3000 psi 400°F	S = 10,300 psi 100°F	S = 10,300 psi 150°F	S = 10,300 psi 200°F	S = 10,300 psi 250°F	S = 10,000 psi 300°F	S = 9700 psi 350°F	S = 9400 psi 400°F
¼	912	775	745	729	714	608	456	1569	1569	1569	1569	1524	1478	1432
⅜	779	662	636	623	610	519	389	1341	1341	1341	1341	1302	1263	1224
½	722	613	589	577	565	481	361	1242	1242	1242	1242	1206	1169	1133
⅝	631	537	516	505	495	421	316	1086	1086	1086	1086	1055	1023	991
¾	582	495	475	466	456	388	291	1002	1002	1002	1002	972	943	914
1	494	420	404	395	387	330	247	850	850	850	850	825	801	776
1¼	439	373	358	351	344	293	219	755	755	755	755	733	711	689
1½	408	347	334	327	320	272	204	702	702	702	702	682	661	641
2	364	309	297	291	285	242	182	625	625	625	625	607	589	570
2½	336	285	274	269	263	224	168	577	577	577	577	560	544	527
3	317	270	259	254	248	211	159	545	545	545	545	529	513	497
3½	304	258	248	243	238	202	152	522	522	522	522	506	491	476
4	293	249	240	235	230	196	147	504	504	504	504	489	474	460

(continued)

(continued)

Nominal Size, in.	Annealed							Drawn*						
	S = 6000 psi 100°F	S = 5100 psi 150°F	S = 4900 psi 200°F	S = 4800 psi 250°F	S = 4700 psi 300°F	S = 4000 psi 350°F	S = 3000 psi 400°F	S = 10,300 psi 100°F	S = 10,300 psi 150°F	S = 10,300 psi 200°F	S = 10,300 psi 250°F	S = 10,000 psi 300°F	S = 9700 psi 350°F	S = 9400 psi 400°F
5	269	229	220	215	211	179	135	462	462	462	462	449	435	422
6	251	213	205	201	196	167	125	431	431	431	431	418	406	393
8	270	230	221	216	212	180	135	464	464	464	464	451	437	424
10	271	231	222	217	212	181	136	466	466	466	466	452	439	425
12	253	215	207	203	199	169	127	435	435	435	435	423	410	397

*When brazing or welding is used to join drawn tube, the corresponding annealed rating must be used.

Courtesy Copper Development Association

Rated Internal Working Pressure for Copper Tube: ACR (Air Conditioning and Refrigeration Field Service)

Tube Size (OD), in.	Annealed							Drawn*						
	COILS													
	S = 6000 psi 100°F	S = 5100 psi 150°F	S = 4900 psi 200°F	S = 4800 psi 250°F	S = 4700 psi 300°F	S = 4000 psi 350°F	S = 3000 psi 400°F	S = 10,300 psi 100°F	S = 10,300 psi 150°F	S = 10,300 psi 200°F	S = 10,300 psi 250°F	S = 10,000 psi 300°F	S = 9700 psi 350°F	S = 9400 psi 400°F
1/8	3074	2613	2510	2459	2408	2049	1537	—	—	—	—	—	—	—
3/16	1935	1645	1581	1548	1516	1290	968	—	—	—	—	—	—	—
1/4	1406	1195	1148	1125	1102	938	703	—	—	—	—	—	—	—
5/16	1197	1017	977	957	937	798	598	—	—	—	—	—	—	—
3/8	984	836	803	787	770	656	492	—	—	—	—	—	—	—
1/2	727	618	594	581	569	485	363	—	—	—	—	—	—	—
5/8	618	525	504	494	484	412	309	—	—	—	—	—	—	—
3/4	511	435	417	409	400	341	256	—	—	—	—	—	—	—
3/4	631	537	516	505	495	421	316	—	—	—	—	—	—	—
7/8	582	495	475	466	456	388	291	—	—	—	—	—	—	—
1 1/8	494	420	404	395	387	330	247	—	—	—	—	—	—	—
1 3/8	439	373	358	351	344	293	219	—	—	—	—	—	—	—
1 5/8	408	347	334	327	320	272	204	—	—	—	—	—	—	—

STRAIGHT LENGTHS

Tube Size (OD), in.	S = 6000 psi 100°F	S = 5100 psi 150°F	S = 4900 psi 200°F	S = 4800 psi 250°F	S = 4700 psi 300°F	S = 4000 psi 350°F	S = 3000 psi 400°F	S = 10,300 psi 100°F	S = 10,300 psi 150°F	S = 10,300 psi 200°F	S = 10,300 psi 250°F	S = 10,000 psi 300°F	S = 9700 psi 350°F	S = 9400 psi 400°F
3/8	914	777	747	731	716	609	457	1569	1569	1569	1569	1524	1478	1432
1/2	781	664	638	625	612	521	391	1341	1341	1341	1341	1302	1263	1224
5/8	723	615	591	579	567	482	362	1242	1242	1242	1242	1206	1169	1133
3/4	633	538	517	506	496	422	316	1086	1086	1086	1086	1055	1023	991
7/8	583	496	477	467	457	389	292	1002	1002	1002	1002	972	943	914
1 1/8	495	421	404	396	388	330	248	850	850	850	850	825	801	776
1 3/8	440	374	359	352	344	293	220	755	755	755	755	733	711	689
1 5/8	409	348	334	327	320	273	205	702	702	702	702	682	661	641
2 1/8	364	309	297	291	285	243	182	625	625	625	625	607	589	570
2 5/8	336	286	275	269	263	224	168	577	577	577	577	560	544	527
3 1/8	317	270	259	254	249	212	159	545	545	545	545	529	513	497
3 5/8	304	258	248	243	238	203	152	522	522	522	522	506	491	476
4 1/8	293	249	240	235	230	196	147	504	504	504	504	489	474	460

*When brazing or welding is used to join drawn tube, the corresponding annealed rating must be used.

Courtesy Copper Development Association

Actual Burst Pressures, Types K, L, and M Copper Water Tube (psi at room temperature)*

Nominal or Standard Size, in.	Actual Outside Diameter, in.	K Drawn	K Annealed	L† Drawn	L† Annealed	M Drawn	M Annealed
½	⅝	9840	4535	7765	3885	6135	—
¾	⅞	9300	4200	5900	2935	4715	—
1	1⅛	7200	3415	5115	2650	3865	—
1¼	1⅜	5525	2800	4550	2400	3875	—
1½	1⅝	5000	2600	4100	2200	3550	—
2	2⅛	3915	2235	3365	1910	2935	—
2½	2⅝	3575	—	3215	—	2800	—
3	3⅛	3450	—	2865	—	2665	—
4	4⅛	3415	—	2865	—	2215	—
5	5⅛	3585	—	2985	—	2490	—
6	6⅛	3425	—	2690	—	2000	—
8	8⅛	3635	—	2650	—	2285	—

*The figures shown are averages of three certified tests performed on each type and size of water tube. In each case, wall thickness was at or near the minimum prescribed for each tube type. No burst pressure in any test deviated from the average by more than 5 percent.

†These burst pressures can be used for ACR tube of equivalent actual OD and wall thickness.

Courtesy Copper Development Association

Pressure Loss of Water Due to Friction in Types K, L and M Copper Tube (psi per linear foot of tube)*

	Nominal on Standard Size, in.†					
Flow,		¼			⅜	
arm	**K**	**L**	***M***	**K**	**L**	***M***
1	0.138	0.118	N/A	0.036	0.023	0.021
2			N/A	0.130	0.084	0.075
3			N/A	0.275	0.177	0.159
4			N/A			
5			N/A			
10			N/A			
15			N/A			
20			N/A			
25			N/A			
30			N/A			
35			N/A			
40			N/A			
45			N/A			
50			N/A			
60			N/A			
70			N/A			
80			N/A			
90			N/A			
100			N/A			
120			N/A			
140			N/A			

**Fluid velocities in excess of 5–8 feet per second are not recommended. Friction loss values shown are for the flow rates that do not exceed a velocity of 8 feet per second.*
†*Italicized friction loss values indicate flow rates that are between 5 feet and 8 feet per second.*
Courtesy Copper Development Association

Pressure Loss of Water Due to Friction in Types K, L and M Copper Tube (psi per linear foot of tube)* *(continued)*

	Nominal on Standard Size, in.†					
Flow,		¼			⅜	
arm	***K***	***L***	***M***	***K***	***L***	***M***
160			N/A			
180			N/A			
200			N/A			
250			N/A			
300			N/A			
350			N/A			
400			N/A			
450			N/A			
500			N/A			
550			N/A			
600			N/A			
650			N/A			
700			N/A			
760			N/A			
1000			N/A			
2000			N/A			

Pressure Loss of Water Due to Friction in Types K, L and M Copper Tube (psi per linear foot of tube)* *(continued)*

	Nominal on Standard Size, in.†					
Flow,		½			¾	
arm	*K*	*L*	*M*	*K*	*L*	*M*
1	0.010	0.008	0.007	0.002	0.001	0.001
2	0.035	0.030	0.024	0.006	0.005	0.004
3	0.074	0.062	0.051	0.014	0.011	0.009
4	0.125	0.106	0.086	0.023	0.018	0.015
5	0.188	0.161	0.130	0.035	0.027	0.023
10				0.126	0.098	0.084
15						
20						
25						
30						
35						
40						
45						
50						
60						
70						
80						
90						
100						
120						
140						
160						
180						
200						
250						

Pressure Loss of Water Due to Friction in Types K, L and M Copper Tube (psi per linear foot of tube)* *(continued)*

	Nominal on Standard Size, in.†					
Flow,		½			¾	
arm	**K**	**L**	**M**	**K**	**L**	**M**
300						
350						
400						
450						
500						
550						
600						
650						
700						
760						
1000						
2000						

Pressure Loss of Water Due to Friction in Types K, L and M Copper Tube (psi per linear foot of tube)* *(continued)*

	Nominal on Standard Size, in.†					
Flow,		**1**			**1¼**	
arm	**K**	**L**	**M**	**K**	**L**	**M**
1	0.000	0.000	0.000	0.000	0.000	0.000
2	0.002	0.001	0.003	0.001	0.000	0.000
3	0.003	0.003	0.001	0.001	0.001	0.001
4	0.006	0.005	0.004	0.002	0.002	0.002
5	0.009	0.007	0.006	0.003	0.003	0.002
10	0.031	0.027	0.023	0.010	0.010	0.009
15	0.065	0.057	0.049	0.022	0.020	0.018
20		0.096	0.084	0.037	0.035	0.031
25				0.057	0.052	0.047
30				0.079	0.073	0.066
35						
40						
45						
50						
60						
70						
80						
90						
100						
120						
140						
160						
180						
200						
250						

Pressure Loss of Water Due to Friction in Types K, L and M Copper Tube (psi per linear foot of tube)* (continued)

	Nominal on Standard Size, in.†					
Flow,		***1***			***1¼***	
arm	**K**	**L**	**M**	**K**	**L**	**M**
300						
350						
400						
450						
500						
550						
600						
650						
700						
760						
1000						
2000						

Pressure Loss of Water Due to Friction in Types K, L and M Copper Tube (psi per linear foot of tube)* *(continued)*

	Nominal on Standard Size, in.†					
Flow,		*1½*			*2*	
arm	*K*	*L*	*M*	*K*	*L*	*M*
1	0.000	0.000	0.000	0.000	0.000	0.000
2	0.000	0.000	0.000	0.000	0.000	0.000
3	0.000	0.000	0.000	0.000	0.000	0.000
4	0.001	0.001	0.001	0.000	0.000	0.000
5	0.001	0.001	0.001	0.000	0.000	0.000
10	0.004	0.004	0.004	0.001	0.001	0.001
15	0.009	0.009	0.008	0.002	0.002	0.002
20	0.016	0.015	0.014	0.004	0.004	0.004
25	0.024	0.022	0.021	0.006	0.006	0.005
30	0.034	0.031	0.29	0.009	0.008	0.008
35	0.045	0.042	0.039	0.012	0.011	0.010
40	0.058	0.054	0.50	0.015	0.014	0.013
45			0.062	0.018	0.017	0.016
50				0.022	0.021	0.020
60				0.031	0.029	0.028
70				0.042	0.039	0.037
80						
90						
100						
120						
140						
160						
180						
200						
250						

Pressure Loss of Water Due to Friction in Types K, L and M Copper Tube (psi per linear foot of tube)* (continued)

	Nominal on Standard Size, in.†					
Flow,		*1½*			*2*	
arm	*K*	*L*	*M*	*K*	*L*	*M*
300						
350						
400						
450						
500						
550						
600						
650						
700						
760						
1000						
2000						

Pressure Loss of Water Due to Friction in Types K, L and M Copper Tube (psi per linear foot of tube)* *(continued)*

	Nominal on Standard Size, in.†					
Flow		***2½***			***3***	
arm	**K**	**L**	***M***	**K**	**L**	***M***
1	0.000	0.000	0.000	0.000	0.000	0.000
2	0.000	0.000	0.000	0.000	0.000	0.000
3	0.000	0.000	0.000	0.000	0.000	0.000
4	0.000	0.000	0.000	0.000	0.000	0.000
5	0.000	0.000	0.000	0.000	0.000	0.000
10	0.000	0.000	0.000	0.000	0.000	0.000
15	0.001	0.001	0.001	0.000	0.000	0.000
20	0.001	0.001	0.001	0.001	0.001	0.001
25	0.002	0.002	0.002	0.001	0.001	0.001
30	0.003	0.003	0.003	0.001	0.001	0.001
35	0.004	0.004	0.004	0.002	0.002	0.001
40	0.005	0.005	0.005	0.002	0.002	0.002
45	0.006	0.006	0.006	0.003	0.003	0.002
50	0.008	0.007	0.007	0.003	0.003	0.003
60	0.011	0.010	0.010	0.005	0.004	0.004
70	0.014	0.014	0.013	0.006	0.006	0.005
80	0.019	0.017	0.016	0.008	0.007	0.007
90	0.023	0.022	0.020	0.010	0.009	0.009
100	0.028	0.026	0.025	0.012	0.011	0.010
120			0.035	0.017	0.016	0.015
140				0.022	0.021	0.019
160				0.028	0.026	0.025
180						
200						
250						

Pressure Loss of Water Due to Friction in Types K, L and M Copper Tube (psi per linear foot of tube)* *(continued)*

	Nominal on Standard Size, in.†					
Flow		**2½**			**3**	
arm	**K**	**L**	**M**	**K**	**L**	**M**
300						
350						
400						
450						
500						
550						
600						
650						
700						
760						
1000						
2000						

Pressure Loss of Water Due to Friction in Types K, L and M Copper Tube (psi per linear foot of tube)* *(continued)*

	Nominal on Standard Size, in.†					
Flow		***4***			***5***	
arm	***K***	***L***	***M***	***K***	***L***	***M***
1	0.000	0.000	0.000	0.000	0.000	0.000
2	0.000	0.000	0.000	0.000	0.000	0.000
3	0.000	0.000	0.000	0.000	0.000	0.000
4	0.000	0.000	0.000	0.000	0.000	0.000
5	0.000	0.000	0.000	0.000	0.000	0.000
10	0.000	0.000	0.000	0.000	0.000	0.000
15	0.000	0.000	0.000	0.000	0.000	0.000
20	0.000	0.000	0.000	0.000	0.000	0.000
25	0.000	0.000	0.000	0.000	0.000	0.000
30	0.000	0.000	0.000	0.000	0.000	0.000
35	0.000	0.000	0.000	0.000	0.000	0.000
40	0.001	0.001	0.000	0.000	0.000	0.000
45	0.001	0.001	0.001	0.000	0.000	0.000
50	0.001	0.001	0.001	0.000	0.000	0.000
60	0.001	0.001	0.001	0.000	0.000	0.000
70	0.002	0.001	0.001	0.001	0.000	0.000
80	0.002	0.002	0.002	0.001	0.001	0.001
90	0.002	0.002	0.002	0.001	0.001	0.001
100	0.003	0.003	0.003	0.001	0.001	0.001
120	0.004	0.004	0.004	0.001	0.001	0.001
140	0.006	0.005	0.005	0.002	0.002	0.002
160	0.007	0.007	0.006	0.002	0.002	0.002
180	0.009	0.008	0.008	0.003	0.003	0.003
200	0.011	0.010	0.010	0.004	0.003	0.003
250	0.016	0.015	0.015	0.006	0.005	0.005

Pressure Loss of Water Due to Friction in Types K, L and M Copper Tube (psi per linear foot of tube)* *(continued)*

	Nominal on Standard Size, in.†					
Flow		4			5	
arm	K	L	M	K	L	M
300			0.021	0.008	0.007	0.007
350				0.010	0.010	0.009
400				0.013	0.012	0.012
450				0.017	0.015	0.015
500						
550						
600						
650						
700						
760						
1000						
2000						

Pressure Loss of Water Due to Friction in Types K, L and M Copper Tube (psi per linear foot of tube)* *(continued)*

	Nominal on Standard Size, in.†					
Flow		*6*			*8*	
arm	*K*	*L*	*M*	*K*	*L*	*M*
1	0.000	0.000	0.000	0.000	0.000	0.000
2	0.000	0.000	0.000	0.000	0.000	0.000
3	0.000	0.000	0.000	0.000	0.000	0.000
4	0.000	0.000	0.000	0.000	0.000	0.000
5	0.000	0.000	0.000	0.000	0.000	0.000
10	0.000	0.000	0.000	0.000	0.000	0.000
15	0.000	0.000	0.000	0.000	0.000	0.000
20	0.000	0.000	0.000	0.000	0.000	0.000
25	0.000	0.000	0.000	0.000	0.000	0.000
30	0.000	0.000	0.000	0.000	0.000	0.000
35	0.000	0.000	0.000	0.000	0.000	0.000
40	0.000	0.000	0.000	0.000	0.000	0.000
45	0.000	0.000	0.000	0.000	0.000	0.000
50	0.000	0.000	0.000	0.000	0.000	0.000
60	0.000	0.000	0.000	0.000	0.000	0.000
70	0.000	0.000	0.000	0.000	0.000	0.000
80	0.000	0.000	0.000	0.000	0.000	0.000
90	0.000	0.000	0.000	0.000	0.000	0.000
100	0.000	0.000	0.000	0.000	0.000	0.000
120	0.001	0.001	0.001	0.000	0.000	0.000
140	0.001	0.001	0.001	0.000	0.000	0.000
160	0.001	0.001	0.001	0.000	0.000	0.000
180	0.001	0.001	0.001	0.000	0.000	0.000
200	0.002	0.001	0.001	0.000	0.000	0.000
250	0.002	0.002	0.002	0.001	0.001	0.001

Pressure Loss of Water Due to Friction in Types K, L and M Copper Tube (psi per linear foot of tube)* *(continued)*

	Nominal on Standard Size, in. [†]					
Flow		***6***			***8***	
arm	***K***	***L***	***M***	***K***	***L***	***M***
300	0.003	0.003	0.003	0.001	0.001	0.001
350	0.004	0.004	0.004	0.001	0.001	0.001
400	0.006	0.005	0.005	0.001	0.001	0.001
450	0.007	0.006	0.006	0.002	0.002	0.002
500	0.008	0.008	0.008	0.002	0.002	0.002
550	0.010	0.009	0.009	0.003	0.002	0.002
600	0.012	0.011	0.011	0.003	0.003	0.003
650		0.013	0.012	0.004	0.003	0.003
700				0.004	0.004	0.004
760				0.006	0.004	0.004
1000				0.008	0.007	0.007
2000						

Pressure Loss of Water Due to Friction in Types K, L and M Copper Tube (psi per linear foot of tube)* *(continued)*

	Nominal on Standard Size, in.†					
Flow		*10*			*12*	
arm	*K*	*L*	*M*	*K*	*L*	*M*
1	0.000	0.000	0.000	0.000	0.000	0.000
2	0.000	0.000	0.000	0.000	0.000	0.000
3	0.000	0.000	0.000	0.000	0.000	0.000
4	0.000	0.000	0.000	0.000	0.000	0.000
5	0.000	0.000	0.000	0.000	0.000	0.000
10	0.000	0.000	0.000	0.000	0.000	0.000
15	0.000	0.000	0.000	0.000	0.000	0.000
20	0.000	0.000	0.000	0.000	0.000	0.000
25	0.000	0.000	0.000	0.000	0.000	0.000
30	0.000	0.000	0.000	0.000	0.000	0.000
35	0.000	0.000	0.000	0.000	0.000	0.000
40	0.000	0.000	0.000	0.000	0.000	0.000
45	0.000	0.000	0.000	0.000	0.000	0.000
50	0.000	0.000	0.000	0.000	0.000	0.000
60	0.000	0.000	0.000	0.000	0.000	0.000
70	0.000	0.000	0.000	0.000	0.000	0.000
80	0.000	0.000	0.000	0.000	0.000	0.000
90	0.000	0.000	0.000	0.000	0.000	0.000
100	0.000	0.000	0.000	0.000	0.000	0.000
120	0.000	0.000	0.000	0.000	0.000	0.000
140	0.000	0.000	0.000	0.000	0.000	0.000
160	0.000	0.000	0.000	0.000	0.000	0.000
180	0.000	0.000	0.000	0.000	0.000	0.000
200	0.000	0.000	0.000	0.000	0.000	0.000
250	0.000	0.000	0.000	0.000	0.000	0.000

Pressure Loss of Water Due to Friction in Types K, L and M Copper Tube (psi per linear foot of tube)* *(continued)*

	Nominal on Standard Size, in.†					
Flow		***10***			***12***	
arm	***K***	***L***	***M***	***K***	***L***	***M***
300	0.000	0.000	0.000	0.000	0.000	0.000
350	0.000	0.000	0.000	0.000	0.000	0.000
400	0.000	0.000	0.000	0.000	0.000	0.000
450	0.001	0.001	0.001	0.000	0.000	0.000
500	0.001	0.001	0.001	0.000	0.000	0.000
550	0.001	0.001	0.001	0.000	0.000	0.000
600	0.001	0.001	0.001	0.000	0.000	0.000
650	0.001	0.001	0.001	0.001	0.000	0.000
700	0.001	0.001	0.001	0.001	0.001	0.001
760	0.002	0.001	0.001	0.001	0.001	0.001
1000	0.003	0.002	0.002	0.001	0.001	0.001
2000				0.004	0.004	0.004

The preceding table is based on the Hazen-Williams formula:

$$P = \frac{4.52Q^{1.85}}{C^{1.85}\, d^{1.87}}$$

Where

P = friction loss, psi per linear foot

Q = flow, gpm

d = average ID, in.

C = constant, 150

Pressure Loss in Fittings and Valves Expressed as Equivalent Length of Copper Tube, ft.*

Nominal or Standard Size, in.	Fittings					Valves			
	Standard Ell		98° Tee						
	90°	45°	Side Branch	Straight Run	Coupling	Ball	Gate	Butterfly	Check
3/8	0.5	—	1.5	—	—	—	—	—	1.5
1/2	1	0.5	2	—	—	—	—	—	2
5/8	1.5	0.5	2	—	—	—	—	—	2.5
3/4	2	0.5	3	—	—	—	—	—	3
1	2.5	1	4.5	—	—	0.5	—	—	4.5
1 1/4	3	1	5.5	0.5	0.5	0.5	—	—	5.5
1 1/2	4	1.5	7	0.5	0.5	0.5	—	—	6.5
2	5.5	2	9	0.5	0.5	0.5	0.5	7.5	9
2 1/2	7	2.5	12	0.5	0.5	—	1	10	11.5
3	9	3.5	15	1	1	—	1.5	15.5	14.5
3 1/2	9	3.5	14	1	1	—	2	—	12.5

4	12.5	5	21	1	1	—	2	16	18.5
5	16	6	27	1.5	1.5	—	3	11.5	23.5
6	19	7	34	2	2	—	3.5	13.5	26.5
8	29	11	50	3	3	—	5	12.5	39

** Allowances are for streamlined soldered fittings and recessed threaded fittings. For threaded fittings, double the allowances shown in the table.*

The equivalent lengths presented above are based upon a C factor of 150 in the Hazen-Williams friction-loss formula. The lengths shown are rounded to the nearest half foot.

Courtesy Copper Development Association

PEX PLASTIC TUBING

Cross-linked polyethylene (PEX) tubing is so named because during the manufacturing process a bridge or link is formed between the polyethylene (PE) macromolecules. The cross-linked PEX molecule results in tubing that resists creep deformation and exhibits durability under temperature extremes and chemical attack.

PEX tubing is marked with the ASTM F 876/F877/CTS-00 SDR9 designation, which means that it was manufactured, inspected, sampled, and tested in accordance with these specifications and was found to meet the specified requirements.

PEX tubing can be used up to 200°F for heating applications. Temperature limitations are always noted on the print line of the tubing. A maximum 140°F operating temperature is recommended.

PEX tubing is commonly available in ¼ through 1 inch CTS (copper tube size), although some tubing manufacturers supply sizes up to a 2-inch diameter. Custom sizes can also be ordered. The pressure ratings are the same for all sizes because the wall thickness is proportionate for each size.

PEX tubing is available in 20-foot straight lengths or coils. Coil tubing is available in a variety of lengths ranging up to a maximum of 1000 feet.

PEX tubing is connected with mechanical fittings. It cannot be joined with solvent cement or heat fusion. The two approved standard specifications for PEX connections are ASTM F 1807 and ASSTM F 1960. The crimp fittings specified in ASTMN F 1807 are the most widely used. Insert and outside diameter compression fittings are also available.

Important performance specifications to consider when using plastic tubing include the following:

- Rated pressure—the short-term burst pressure at 75°F
- Maximum rated vacuum—most frequently given in inches or mm of mercury referenced below one standard atmosphere
- Minimum bend radius—based on acceptable tubing cross-section deformation
- Temperature range—the full required range of ambient operating temperature

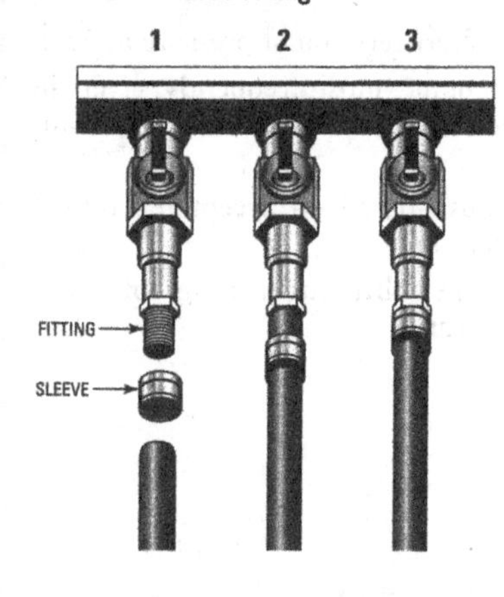

Crimping Fittings

1. Expand the end of the PEX tubing with the expansion tool provided by the PEX tube manufacturer.
2. Insert the brass fitting into the end of the expanded PEX tube.
3. Use the expansion tool to pull the brass sleeve back over the PEX tube and fitting for a tight connection.

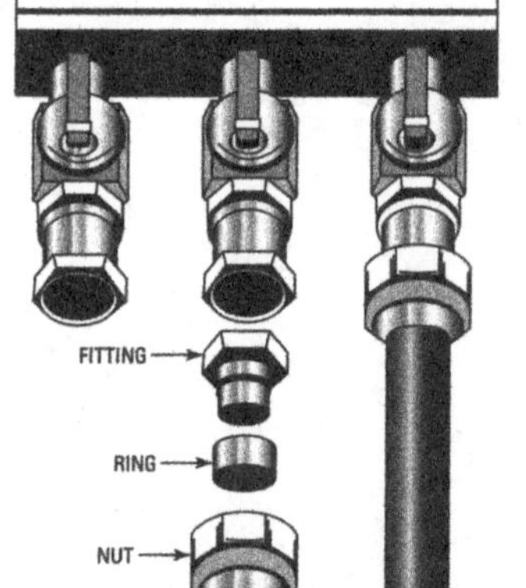

Compression Fitting

1. Slide the locking nut and split compression ring up the tubing.
2. Insert the tubing onto the compression fitting.
3. Tighten the nut onto the compression fitting snugly.
4. Re-tighten the fittings after the heat has been turned on and the hot water has circulated through the tubing.

PEX tubing fittings.

(Courtesy Watts Radiant, Inc.)

Properties of Vanex Cross-Linked Polyethylene (PEX) Tubing

Property	ASTM Test Method	Typical Values	
		English Units	SI Units
Density	D 792	—	0.946 g/cc
Melt index* (190°C/2.16 kg)	D 1238	—	0.7g/10 min
Flexural modulus*	D 790	120,000 psi	830 MPa
Tensile strength at yield (2 in./min)	D 638	2,900 psi	20 MPa
Coefficient of linear thermal expansion at 68°F	D 696	8×10^{-5}°F	15×10^{-5}°C
Hydrostatic design basis			
at 73°F (23°C)	D 2837	1250 psi	8.6 MPa
at 180°F (82°C)	D 2837	800 psi	5.5 MPa
Vicat softening point	D 1525	255°F	124°C
Thermal Conductivity	C 177	2.4 Btu/(h)(ft^2)(°F/in.)	$3.5 \times 10^{\circ 3}$ W/(cm^2)(°C/cm)

*Before Cross-linking

†73° F

Courtesy Vanguard Piping Systems, Inc.

Vanex Cross-Linked Polyethylene (PEX) Tubing Dimensions

SDR9 PEX Tubing ASTM F 876/F 877/CTS-OD SDR9*

Part Number	*Tubing Size*	*OD*	*Wall Thickness*	*Nom. I.D.*	*Weight Per Foot*	*Volume (Gal.) Per 100 Foot*
PX1	¼	0.375±.002	0.065±.002	0.250	0.0261	0.25
PX2	⅜	0.500±.002	0.075±.002	0.350	0.0413	0.50
PX3	½	0.625±.002	0.075±.002	0.475	0.0535	0.92
PX58	⅝	0.750±.002	0.088±.002	0.574	0.0752	1.34
PX4	¾	0.875±.002	0.102±.002	0.671	0.1023	1.82
PX5	1"	1.125±.003	0.132±.002	0.863	0.1689	3.04
PX6	1¼	1.375±.003	0.161±.003	1.053	0.2523	4.52
PX7	1½	1.625±.004	0.191±.004	1.243	0.3536	6.30
PX8	2	2.125±.004	0.248±.004	1.629	0.6010	10.83

** Dimensions are in English units. Tolerances shown are ASTM requirements. For Red or Blue part numbers, place an "R" after the "X" in the part number for the Red tubing and a "B" for the Blue tubing.*

Courtesy Vanguard Piping Systems, Inc.

Pressure Drop in Vanex Cross-Linked Polyethylene (PEX) Tubing

gpm	Size, in.							
	3/8"	½"	5/8"	3/4"	1"	1¼"	1½"	2"
1	0.70	0.016						
1.5	0.149	0.034						
2	0.254	0.058	0.023					
2.5	0.381	0.087	0.035					
3	0.539	0.122	0.048	0.023				
3.5	0.717	0.162	0.065	0.030				
4		0.207	0.083	0.039				
5		0.314	0.125	0.059				
6		0.440	0.175	0.082	0.024			
7		0.586	0.233	0.109	0.032			
8			0.298	0.140	0.041			
9			0.371	0.173	0.051			
10			0.451	0.211	0.062	0.023		
11				0.252	0.074	0.028		
12				0.296	0.087	0.033		
13				0.343	0.101	0.038		

(continued)

gpm	Size, in.							
	3/8"	½"	5/8"	3/4"	1"	1¼"	1½"	2"
14					0.116	0.044		
16					0.148	0.056	0.025	
18					0.184	0.069	0.031	
20					0.224	0.084	0.038	
22					0.267	0.101	0.045	
24						0.118	0.053	
26						0.137	0.061	
28						0.157	0.070	
30						0.179	0.080	0.021
32						0.202	0.090	0.024
34							0.101	0.027
36							0.112	0.030
38							0.124	0.033
40							0.136	0.036
45								0.045
50								0.055

(continued)

gpm	Size, in.							
	3/8"	½"	5/8"	3/4"	1"	1¼"	1½"	2"
55								0.066
60								0.077
65								0.089
70								0.103
75								0.116

**Expressed as psi/ft. pressure drop. Indicates 8 tps maximum velocity required by some plumbing codes.*
Maximum now for each size based on 12 FPS velocity.
PSI × 2.307 = head loss.
Courtesy Vanguard Piping Systems, Inc.

Minimum Burst Pressure (psi) for PEX Tubing (per ASTM F 876/F 877)

Size, in.	*73°F (23°C)*	*180°F (82°C)*
1/4	870	390
3/8	620	275
1/2	480	215
5/8	475	210
3/4	475	210
1	475	210
1 1/4	475	210
1 1/2	475	210
2	475	210

Courtesy Vanguard Piping Systems, Inc.

Minimum Bending Radius for PEX Tubing

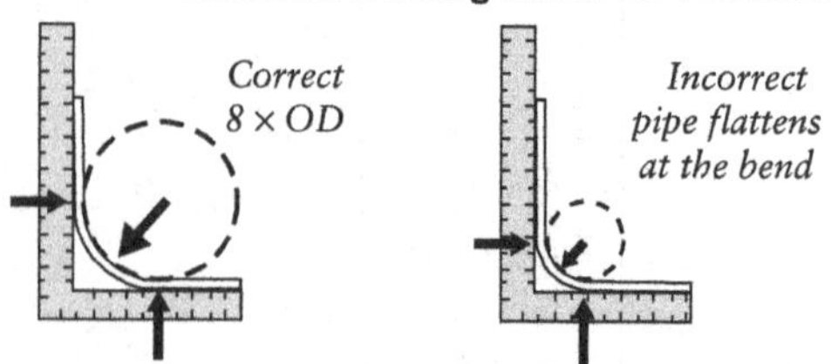

Changes in direction may be made by bending PEX tubing. The minimum bend radius is 8 times the outside diameter. When bending against the coil direction, the minimum bend radius is 24 times the O/D. No special tools are necessary.

Courtesy Vanguard Piping Systems, Inc.

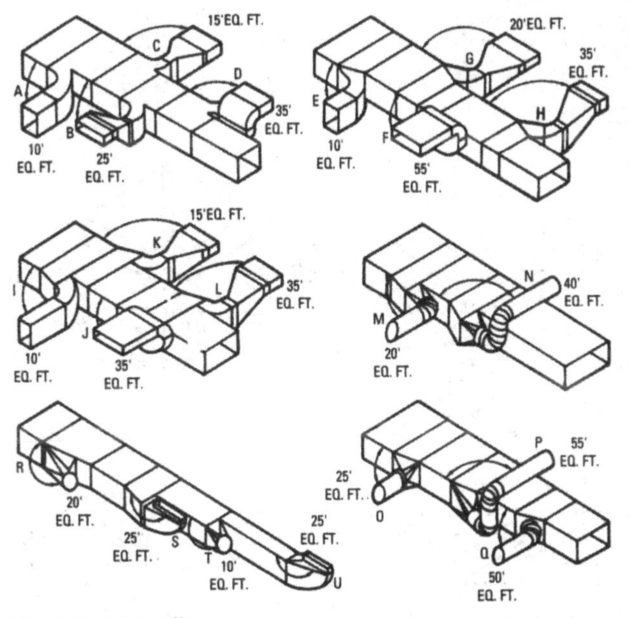

Trunk duct takeoff.

(Courtesy ASHRAE 1952 Guide)

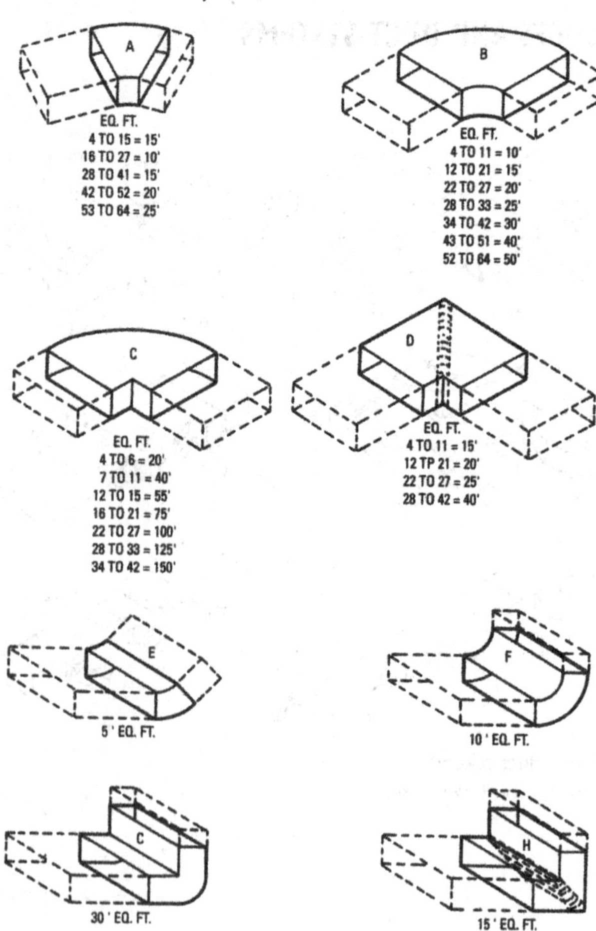

Angles and elbows for trunk ducts.
(Courtesy ASHRAE 1952 Guide)

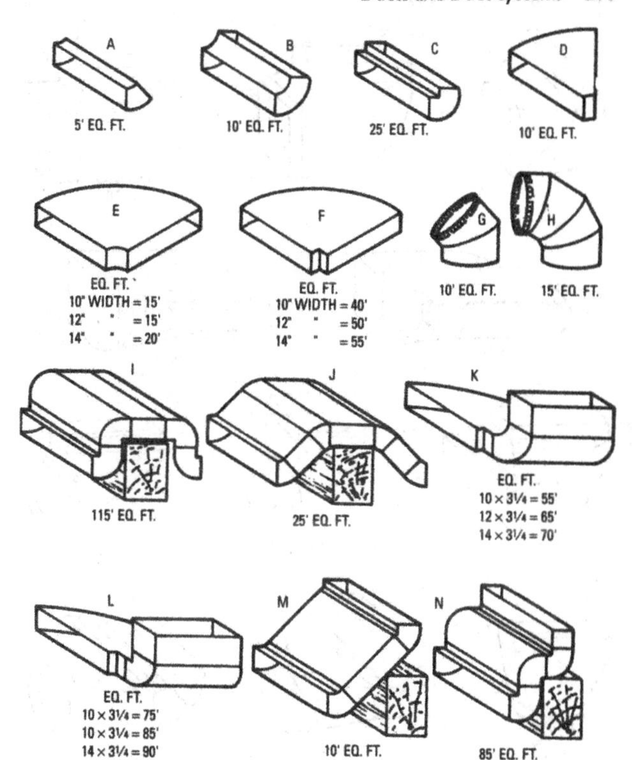

Stock angles and elbows.

(Courtesy ASHRAE 1952 Guide)

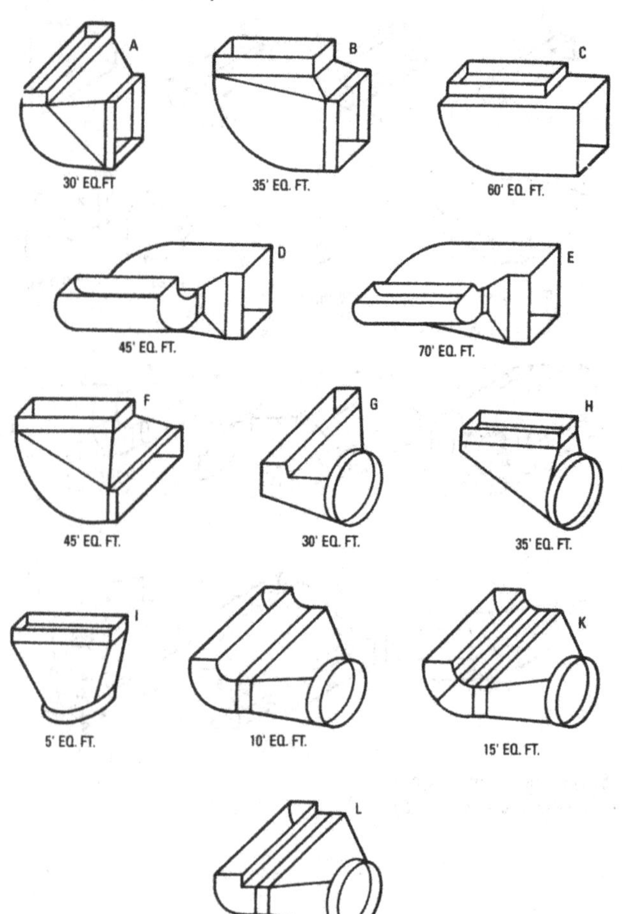

Boot fittings from branch to stack.

(Courtesy ASHRAE 1952 Guide)

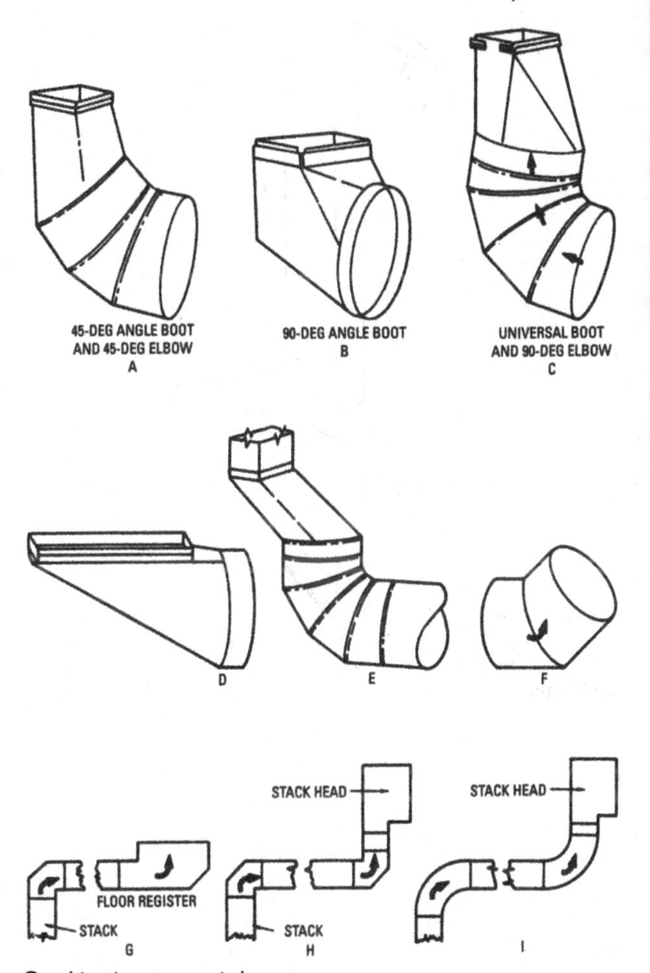

Combination warm-air boots.

(Courtesy ASHRAE 1952 Guide)

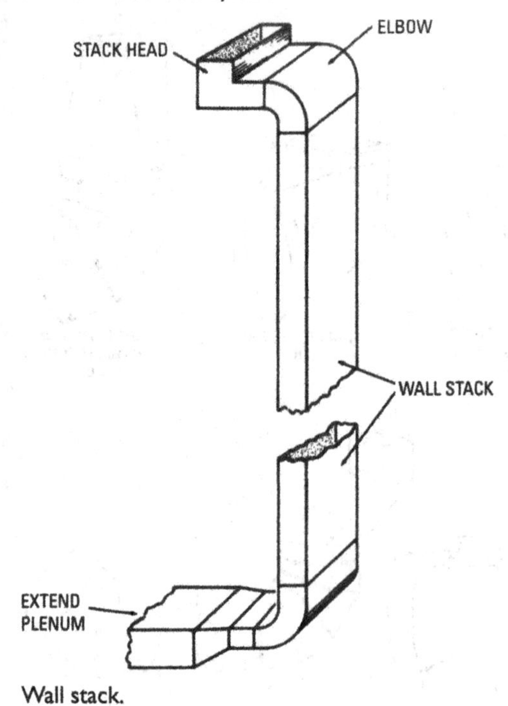

Wall stack.

Recommended and Maximum Air Velocities

Designation	Residences	Schools, Theaters, Public Buildings	Industrial Buildings
	Recommended Velocities (fpm)		
Outdoor air intakes*	500	500	500
Filters*	250	300	350
Heating coils*	450	500	600
Air washers	500	500	500
Fan outlets	1000–1600	1300–2000	1600–2400
Main ducts	700–900	1000–1300	1200–1800
Branch ducts	600	600–900	800–1000
Branch risers	500	600–700	800
	Maximum Velocities (fpm)		
Outdoor air intakes*	800	900	1200
Filters*	300	350	350
Heating coils*	500	600	700
Air washers	500	500	500
Fan outlets	1700	1500–2200	1700–2800
Main ducts	800–1200	1100–1600	1300–2200
Branch ducts	700–1000	800–1300	1000–1800
Branch risers	650–800	800–1200	1000–1600

**These velocities are for total face area, not the net free area; other velocities in table are for net free area.*
(Courtesy ASHRAE 1960 Guide)

Thicknesses, Gauges, and Weights of Plain (Black) and Galvanized Sheet Metal

	U.S. Std. Gauge	*Approximate Thickness (in)*		*Weight per Square Foot*	
		Steel	*Iron*	*Ounces*	*Pounds*
Black sheets	30	0.0123	0.0125	8	0.500
	28	0.0153	0.0156	10	0.625
	26	0.0184	0.0188	12	0.750
	24	0.0245	0.0250	16	1.000
	22	0.0306	0.0313	20	1.250
	20	0.0368	0.0375	24	1.500
	18	0.0490	0.0500	32	2.000
	16	0.0613	0.0625	40	2.500
	14	0.0766	0.0781	50	3.125
	12	0.1072	0.1094	70	4.375
	11	0.1225	0.1250	80	5.000
	10	0.1379	0.1406	90	5.625
Galvanized sheets*	30	0.0163	0.0165	10.5	0.656
	28	0.0193	0.0196	12.5	0.781
	26	0.0224	0.0228	14.5	0.906
	24	0.0285	0.0290	18.5	1.156
	22	0.0346	0.0353	22.5	1.406
	20	0.0408	0.0415	26.5	1.656
	18	0.0530	0.0540	34.5	2.156
	16	0.0653	0.0665	42.5	2.656
	14	0.0806	0.0821	52.5	3.281
	12	0.1112	0.1134	72.5	4.531
	11	0.1265	0.1290	82.5	5.156
	10	0.1419	0.1446	92.5	5.781

** Galvanized sheets are gauged before galvanizing and are therefore approximately 0.004 inch thicker. (Courtesy ASHRAE 1960 Guide)*

Thicknesses, Gauges, and Weights of 2S Aluminum (density 0.098 lb/in³)

B. & S.	Thickness (in)		Weight per Square Foot	
Gauge	Decimal	Nearest Fraction	Pounds	Ounces
28	0.012	1/64	2.7	0.169
26	0.016	1/64	3.6	0.226
24	0.020	1/64	4.5	0.282
22	0.025	1/32	5.4	0.353
20	0.032	1/32	7.2	0.452
18	0.040	3/64	9.0	0.563
16	0.051	3/64	11.5	0.720
14	0.064	1/16	14.4	0.903

(Courtesy ASHRAE 1960 Guide)

Equivalents of Round Duct Diameters

Duct Diameter, in.	*Duct Diameter, mm*	*Area, Ft²*	*Area, m²*
8	203	0.3491	0.032
10	254	0.5454	0.051
12	305	0.7854	0.073
14	356	1.069	0.099
16	406	1.396	0.130
18	457	1.767	0.290
20	508	2.182	0.203
22	559	2.640	0.245
24	609	3.142	2.292
26	660	3.687	0.342
28	711	4.276	0.397
30	762	4.900	0.455
32	813	5.585	0.519
34	864	6.305	0.586
36	914	7.069	0.657
38	965	7.786	0.732
40	1016	8.727	0.811
42	1067	9.62	0.894
44	1119	10.56	0.981
46	1168	11.54	1.072
48	1219	12.57	1.168
50	1270	13.67	1.270
52	1321	14.75	1.370
54	1372	15.90	1.477
56	1422	17.10	1.586
58	1473	18.35	1.705
60	1524	19.63	1.824

Natural Gas Manifold Pressures for Gas-Fired Duct Furnaces

Btu per Cubic Foot	*Sp. Gr.*	*Man. Press. (Inches of Water)*	*Btu per Cubic Foot*	*Sp. Gr.*	*Man. Press. (Inches of Water)*
900	0.50	3.4	1000	0.55	3.0
	0.55	3.7		0.60	3.3
	0.60	4.1		0.65	3.6
	0.65	4.4		0.70	3.9
925	0.50	3.2	1025	0.55	2.9
	0.55	3.5		0.60	3.1
	0.60	3.9		0.65	3.4
	0.65	4.2		0.70	3.7
950	0.50	3.1	1050	0.55	2.7
	0.55	3.4		0.60	3.0
	0.60	3.7		0.65	3.2
	0.65	4.0		0.70	3.5
	0.70	4.3			
975	0.50	3.2	1075	0.55	2.6
	0.60	3.5		0.60	2.9
	0.65	3.8		0.65	3.2
	0.70	4.1		0.70	3.3
			1100	0.55	2.5
				0.60	2.7
				0.65	2.9
				0.70	3.2

Note: Manifold pressures on this table are based on orifice sizes as shown in orifice table in installation manual. Pressures given in this table apply to sizes of units. This table does not apply to units used in high-altitude areas. See supplement for high-altitude manifold pressure table.

(Courtesy Janitrol)

Maximum Workload Data for Damper Motors

Connection Position	1		2		3		4		5	
Length of Travel (in)	3.25*		3.00*		2.75*		2.50*		2.25*	
Specifications	Max Load (lbs)	Return Force (lbs)	Max Load (lbs)	Return Force (lbs)	Max Load (lbs)	Return Force (lbs)	Max Load (lbs)	Return Force (lbs)	Max Load (lbs)	Return Force (lbs)
25 VA	15	8½	17	9½	19	10	20	11	22	13
40 VA	30	8½	34	9½	37	10	42	11	45	13
120 VA	30	8½	34	9½	37	10	42	11	45	13
Length of Travel (in)	2.625†		2.375†		2.125†		1.875†		1.625†	
25 VA	18	10	21	11	24	13	27	15	30	17
40 VA	36	10	42	11	48	13	54	15	60	17
120 VA	36	10	42	11	48	13	54	15	60	17

** When using dual damper arm, combined load must not exceed load maximums shown.*
† Table for reversed damper arm position.
(Courtesy ITT General Controls)

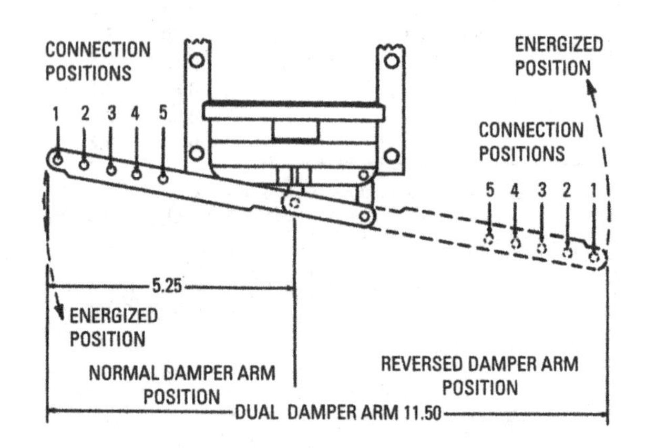

THERMOSTATS

Terminal Identification and Color Coding*

Terminal	*Proposed NEMA Standard*	*Color*	*Comments*
R or V	24 VAC power	Red	—
Rh or 4	24 VAC heating power	Red	—
Rc	24 VAC cooling power	Red	—
C	24 VAC common	Black	—
X	24 VAC common	Normally black, but other colors also used	Used by some manufacturers for the emergency heat relay
X2	Second stage heating or indicator lights on some thermostats	No set standard color	Used by some manufacturers for the emergency heat relay
Y	Cooling/ compressor contactor	Yellow	Cooling or first stage heating on a heat pump
Y2	2-stage cooling	Blue or orange	None

(continued)

(continued)

Terminal	*Proposed NEMA Standard*	*Color*	*Comments*
W	Heating	White	First-stage heating may require a jumper to Y on a heat pump. Second-stage heating on some systems
W2	Second-stage heating	No set standard color	First-stage auxiliary heating on a heat pump
G	Fan	Green	Fan switch on thermostat or on a call for cooling or heat pump
E	Emergency heat relay	No set standard color	Used to disable the heat pump and make auxiliary heating first stage
O	Reversing valve cooling	No set standard color	None
B	Heating change-over valve	No set standard color	Used by some manufacturers for the common side of the transformer
T	Outdoor anticipator	No set standard color	None

(continued)

Terminal	Proposed NEMA Standard	Color	Comments
L	Service light	No set standard color	Used to notify you of a problem with the system

** There is no common standard for thermostat wire colors or terminal identification. Always closely follow the thermostat manufacturer's wiring diagram when installing a new unit.*

Anticipator Values

Amperes	Color	Amperes	Color
0.83–0.72	Brown-red	0.33–0.29	Orange-yellow
0.72–0.68	Brown-blue	0.29–0.25	Orange-green
0.68–0.55	Orange	0.25–0.22	Red
0.55–0.48	Blue	0.22–0.19	Green-blue
0.48–0.41	Blue-orange	0.11–0.10	Orange-red
0.41–0.36	Blue-yellow	0.10–0.09	Red-yellow
0.36–0.33	Green	0.09–0.08	Green-yellow

(Courtesy ITT General Controls)

HUMIDIFIERS AND DEHUMIDIFIERS

Air that has a low relative humidity will absorb water vapor from any available source. Most important from a personal standpoint is the evaporation of moisture from the membranes of the nose, mouth, and throat. These are our protective zones, and excessive dryness of these membranes will cause discomfort.

It is difficult to pinpoint the most desirable level of relative humidity, but it is generally agreed that the range between 30 and 50 percent is the best from both health and comfort standpoints. The upper part of this range (40 to 50 percent) is impractical on very cold days because of condensation on windows. Therefore, it is recommended that a relative humidity between 30 and 40 percent be maintained during the heating season.

Relationship Between Temperature and Relative Humidity

Outside Temperature	*Outside Relative Humidity (%)*	*Inside Temperature*	*Inside Relative Humidity (%)*
	0		0
	20		3
20°F	40	70°F	6
	60		8
	80		11

Recommended Scale of Interior Effective Temperatures for Various Outside Dry-Bulb Conditions

Degrees Outside	*Degrees Inside*			
Dry-Bulb	***Dry-Bulb***	***Wet-Bulb***	***Dew Point***	***Effective Temperature***
100	82.5	69.0	62.3	76.0
95	81.0	67.7	60.8	74.8
90	79.5	66.5	59.5	73.6
85	78.1	65.3	58.0	72.5
80	76.7	64.0	56.6	71.3
75	75.3	63.0	55.6	70.2
70	74.0	62.0	54.5	69.0

The *temperature-humidity* index (formerly called the *discomfort index*) is a numerical indicator of human discomfort caused by temperature and moisture. It is calculated by adding the indoor dry-bulb and the indoor wet-bulb readings, multiplying the sum by 0.4 and adding 15.

COMPRESSORS AND CONDENSERS

Compressor Size and Motor Current Ratings

Single-Phase Motor		*Three-Phase Motor*	
Compressor Size, tons	***Motor Current, amperes***	***Compressor Size, tons***	***Motor Current, amperes***
2	18	3	18
3	25–30	4	25–30
4	30–40	5	30–40
5	35–50	7½	35–50

Courtesy Honeywell Tradeline Controls

Compressor Characteristics

Compressor Performance. The ability of a compressor to provide the maximum refrigeration effect with the least amount of power input. The performance factor of a compressor may be written as follows:

Performance factor = capacity (Btu)/
power input (watts)

Compressor Capacity. Useful refrigeration effect. It is equal to the difference in the total enthalpy between the refrigerant liquid at a temperature corresponding to the pressure leaving the compressor and the refrigerant vapor entering the compressor measured in Btus.

Compressor Mechanical Efficiency. The ratio of the work delivered to the gas (as obtained by an indicator diagram) to the work delivered to the compressor shaft.

Compressor Volumetric Efficiency. The ratio of the actual volume of a refrigerant gas pumped by a compressor to the volume displaced by the compressor piston.

Compressor Compression Ratio. The ratio of the absolute head pressure to the absolute suction pressure.

REFRIGERANTS

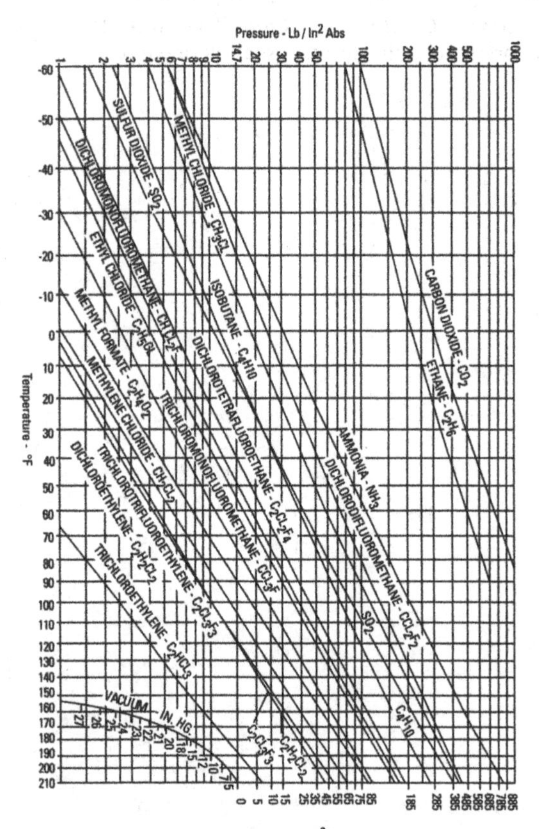

Vapor pressure of common refrigerants at various temperature ranges.

The figure is a logarithmic-scaled chart giving the relation between pressure and corresponding temperatures in degrees Fahrenheit for common refrigerants. The axis of the abscissa shows the temperature, and the axis of the ordinate shows the pressure in psi and absolute, respectively.

To ascertain the pressure of a refrigerant at any particular temperature, follow the desired temperature until the curve of that particular refrigerant is reached. The corresponding pressure is found on the pressure axis. For example, the temperature corresponding to a pressure of 25 psi for sulfur dioxide is approximately 60°F, and the corresponding temperature at the same pressure for dichlorodifluoromethane (freon-12) is approximately 27.2°F.

Characteristics of Typical Refrigerants

*Name**	*Boiling point, °F*	*Heat of Vaporization at Boiling Point, Btu/lb (at 1 atm)*
Sulfur dioxide (SO_2)	14.0	172.3
Methyl chloride (CH_3Cl)*	−10.6	177.80
Ethyl chloride (C_2H_5Cl)	55.6	177.00
Ammonia (NH_3)	−28.0	554.70
Carbon dioxide (CO_2)	−110.5	116.00
Freezol (isobutane) [(CH_3)3CH]	10.0	173.50
Freon-11 (CCl_3F)*	74.8	78.31
Freon-12 (CCl_2F)*	−21.7	71.04
Freon-13 ($CClF_3$)*	−114.6	63.85
Freon-21 ($CHCl_2F$)*	48.0	104.15
Freon-22 ($CHClF_2$)*	−41.4	100.45
Freon-113 ($CCl_2F—CClF_2$)*	117.6	63.12
Freon-114 ($CClF_2$)*	38.4	58.53
Freon-115 ($CClF_2CF_3$)*	−37.7	54.20
Freon-502	−50.1	76.46

**Note:* The freon family of refrigerants, originally designated as freon-12, freon-13, etc., are presently listed under various tradenames, such as Ucon-12, Ucon-22, or simply Refrigerant-12, Refrigerant-22, sometimes abbreviated R-12, R-22, R-113, etc., depending on the particular refrigerant characteristics desired.

Refrigerant Pressure versus Temperature

	Pressure at Sea Level (pai) Gage			
Temperature, °F	R-12	R-22	R502	R717 (NH_3)
−40	11.0*	0.5	4.1	8.7*
−35	8.4*	2.6	6.5	5.4*
−30	5.5*	4.9	9.2	1.6*
−25	2.3*	7.4	12.1	1.3
−20	0.6*	10.1	15.3	3.6
−15	2.4	13.2	18.8	6.2
−10	4.5	16.5	22.6	9.0
−5	6.7	20.1	26.7	12.2
0	9.2	24.0	31.1	15.7
5	11.8	28.2	35.9	18.6
10	14.6	32.8	41.0	23.8
15	17.7	37.7	46.5	28.4
20	21.0	43.0	52.5	33.5
25	24.6	48.8	58.8	39.0
30	28.5	54.9	65.6	45.0
35	32.6	61.5	72.8	51.6
40	37.0	68.5	80.5	58.6
45	41.7	76.0	88.7	66.3
50	46.7	84.0	97.4	74.5
55	52.0	92.6	106.6	83.4
60	57.7	101.6	116.4	92.9
65	63.8	111.2	126.7	103.1
70	70.2	121.4	137.6	114.1
75	77.0	132.2	149.1	125.8
80	84.2	143.6	161.2	138.3
85	91.8	155.7	174.0	151.7
90	99.8	168.4	187.4	165.9
95	108.2	181.8	201.4	181.1
100	117.2	195.9	216.2	197.2
105	126.6	210.8	231.7	214.2
110	136.4	226.4	247.9	232.3
115	146.8	242.7	264.9	251.5
120	157.6	259.9	282.7	271.7
125	169.1	277.9	301.4	293.1
130	181.0	296.8	320.8	
135	193.5	316.6	341.3	
140	206.6	337.2	362.6	

*In Hg (mercury) standard atmosphere.

Quantities of Refrigerant Circulated per Minute Under Standard Ton Conditions

Refrigerant	*Pounds Expanded per Minute*	*Ft³/lb Liquid 86°F*	*In³ Liquid Expanded per Minute*	*Specific Gravity Liquid 86°F (Water-1)*
Carbon dioxide	3.528	0.0267	162.8	0.602
Freon-22	2.887	0.01367	67.97	1.177
Ammonia	0.4215	0.02691	19.6	0.598
Freon-12	3.916	0.0124	83.9	1.297
Methyl chloride	1.331	0.01778	40.89	0.898
Sulfur dioxide	1.414	0.01184	28.9	1.358
Freon-114	4.64	0.01112	89.16	1.443
Freon-21	2.237	0.01183	45.73	1.360
Freon-11	2.961	0.01094	55.976	1.468
Methylene chloride	1.492	0.01198	30.88	1.340
Freon-113	3.726	0.01031	66.48	1.555

$$\text{Cu in. liquid refrig./min} = \frac{200\ \text{Btu/min}}{\text{Btu refrig. effect/lb}} \times \text{vol. liquid/lb } 86°\text{F} \times 1728$$

Factors to Be Added to Initial Temperature of Coolant to Determine Condenser Temperature

Evap. °F	Initial Coolant Temp. °F				
	60°	*70°*	*80°*	*90°*	*100°*
–30	15	15	15	10	10
–25	15	15	15	15	10
–20	20	20	15	15	15
–15	20	20	20	15	15
–10	20	20	20	15	15
– 5	20	20	20	15	15
0	25	25	20	20	15
+5	30	30	30	25	20
+10	35	35	30	25	20
+15	40	35	30	25	20
+20	40	35	30	25	25
+25	45	40	35	30	30
+30	50	45	40	35	35
+35	50	50	45	45	40

Pressure in Pounds Per Square Inch (Gauge) or Inches of Vacuum Corresponding to Temperature in Degrees F for Various Common Refrigerants

Temp. °F	Ammonia	Sulfur Dioxide	Methyl Chloride	Ethane	Propane	Ethyl Chloride	Carbon Dioxide	Butane	Isobutane	Freon F-12	Correne	Methyl Formate
–40	8.7 in	23.5 in	15.7 in	99.8 psig	1.5 psig		131.1 psig			11.0 in		
–35	5.4 in	22.4 in	14.4 in	109.8 psig	3.4 psig		156.3 psig			8.4 in		
–30	1.6 in	21.1 in	11.6 in	120.3 psig	5.6 psig		163.1 psig			5.5 in		
–25	1.3 psig	19.6 in	9.2 in	132.0 psig	8.0 psig		176.3 psig			2.3 in		
–20	3.6 psig	17.9 in	6.1 in	144.8 psig	10.7 psig	25.3 in	205.8 psig		14.6 in	0.5 psig		
–15	6.2 psig	16.1 in	2.3 in	157 psig	13.6 psig	24.5 in	225.8 psig		13.0 in	2.4 psig		
–10	9.0 psig	13.9 in	0.2 psig	172 psig	16.7 psig	23.6 in	247.0 psig		11.0 in	4.5 psig	28.1 in	
–5	12.2 psig	11.5 in	2.0 psig	187 psig	20.0 psig	22.6 in	269.7 in		8.8 in	6.8 psig	27.8 in	
0	15.7 psig	8.8 in	3.8 psig	204 psig	23.5 psig	21.5 in	293.9 psig	15.0 in	6.3 in	9.2 psig	27.5 in	26.5 in
+5	19.6 psig	5.8 in	6.2 psig	221 psig	27.4 psig	20.3 in	319.7 psig	12.2 in	3.3 in	11.9 psig	27.1 in	25.9 in
+10	23.8 psig	2.6 in	8.6 psig	239 psig	31.4 psig	18.9 in	347.1 psig	11.1 in	0.2 in	14.7 psig	26.7 in	25.4 in
+15	28.4 psig	0.5 psig	11.2 psig	257 psig	35.9 psig	17.4 in	376.3 psig	8.8 in	1.6 psig	17.7 psig	26.2 in	24.7 in
+20	33.5 psig	2.4 psig	13.6 psig	275 psig	40.8 psig	15.8 in	407.3 psig	6.3 in	3.5 psig	21.1 psig	25.6 in	24.0 in
+25	39.0 psig	4.6 psig	17.2 psig	292 psig	46.2 psig	14.0 in	440.1 psig	3.6 in	5.5 psig	24.6 psig	24.9 in	23.1 in
+30	45.0 psig	7.0 psig	20.3 psig	320 psig	51.6 psig	12.2 in	474.9 psig	0.6 in	7.6 psig	28.5 psig	24.3 in	22.3 in
+35	51.6 psig	9.6 psig	24.0 psig	343 psig	57.3 psig	10.1 in	511.7 psig	1.3 psig	9.9 psig	32.6 psig	23.5 in	21.1 in

+40	58.6 psig	12.4 psig	28.1 psig	368 psig	63.3 psig	8.0 in	550.7 psig	3.0 psig	12.2 psig	37.0 psig	22.6 in	20.0 in
+45	66.3 psig	15.5 psig	32.2 psig	390 psig	69.9 psig	5.4 in	591.8 psig	4.9 psig	14.8 psig	41.7 psig	21.7 in	18.7 in
+50	74.5 psig	18.8 psig	36.3 psig	413 psig	77.1 psig	2.3 in	635.3 psig	6.9 psig	17.8 psig	46.7 psig	20.7 in	17.3 in
+55	83.4 psig	22.4 psig	41.7 psig	438 psig	84.6 psig	0.3 psig	681.2 psig	9.1 psig	20.8 psig	52.0 psig	19.5 in	15.7 in
+60	92.9 psig	26.2 psig	46.3 psig	466 psig	92.4 psig	1.9 psig	729.5 psig	11.6 psig	24.0 psig	57.7 psig	18.2 in	14.0 in
+65	103.1 psig	30.4 psig	53.6 psig	496 psig	100.7 psig	3.3 psig	780.4 psig	14.2 psig	27.5 psig	63.7 psig	16.7 in	11.9 in
+70	114.1 psig	34.9 psig	57.8 psig	528 psig	109.3 psig	6.2 psig	834.0 psig	16.9 psig	31.1 psig	70.1 psig	15.1 in	9.8 in
+75	125.8 psig	39.8 psig	64.4 psig	569 psig	118.5 psig	8.3 psig	890.4 psig	19.8 psig	35.0 psig	76.9 psig	13.4 in	7.3 in
+80	138.3 psig	45.0 psig	72.3 psig	610 psig	128.1 psig	10.5 psig	949.6 psig	22.9 psig	39.2 psig	84.1 psig	11.5 in	4.9 in
+85	151.7 psig	50.9 psig	79.4 psig	657 psig	138.4 psig	12.9 psig	1011.3 psig	26.2 psig	43.9 psig	91.7 psig	8.4 in	2.4 in
+90	165.9 psig	56.5 psig	87.3 psig	693 psig	149 psig	15.4 psig		29.8 psig	48.6 psig	99.6 psig	7.3 in	0.5 psig
+95	181.1 psig	62.9 psig	95.6 psig		160 psig	18.2 psig		33.2 psig	53.7 psig	108.1 psig	5.0 in	2.1 psig
+100	197.2 psig	69.8 psig	102.3 psig		172 psig	21.0 psig		37.5 psig	59.0 psig	116.9 psig	2.4 in	3.8 psig
+105	214.2 psig	77.1 psig	113.4 psig		185 psig	24.3 psig		41.7 psig	64.6 psig	126.2 psig	0.19 psig	5.8 psig
+110	232.2 psig	85.1 psig	118.3 psig		197 psig	27.3 psig		46.1 psig	70.4 psig	136.0 psig	1.6 psig	7.7 psig
+115	251.5 psig	93.5 psig	128.6 psig		207.6 psig	31.6 psig			76.7 psig	146.5 psig	3.1 psig	10.4 psig
+120	271.7 psig	106.4 psig	139.3 psig		218.3 psig	35.5 psig			84.3 psig	157.1 psig	4.7 psig	13.1 psig
+125	293.1 psig	111.9 psig	150.3 psig		232.3 psig	39.5 psig			90.1 psig	168.6 psig	6.6 psig	15.7 psig
+130	315.6 psig	121.9 psig	161.3 psig		246.3 psig	44.0 psig			97.3 psig	180.2 psig	8.4 psig	18.2 psig

Note: in = inches of mercury

VENTILATION AND EXHAUST FANS

Fresh Air Change Requirements—Minimum Air Changes per Hour

Type of Building or Room	*Minimum Air Changes per Hour*	*Cubic Feet of Air per Minute per Occupant*
Attic spaces (for cooling)	12–15	
Boiler room	15–20	
Churches, auditoriums	8	20–30
College classrooms		25–30
Dining rooms (hotel)	5	
Engine rooms	4–6	
Factory buildings (ordinary manufacturing)	2–4	
Factory buildings (extreme fumes or moisture)	10–15	
Foundries	15–20	

(continued)

(continued)

Type of Building or Room	*Minimum Air Changes per Hour*	*Cubic Feet of Air per Minute per Occupant*
Galvanizing plants	20–30	
Garages (repair)	20–30	
Garages (storage)	4–6	
Homes (night cooling)	9–17	
Hospitals (general)		40–50
Hospitals (children's)		35–40
Hospitals (contagious diseases)		80–90
Kitchens (hotel)	10–20	
Kitchens (restaurant)	10–20	
Libraries (public)	4	
Laundries	10–15	
Mills (paper)	15–20	
Mills (textile—general buildings)	4	
Mills (textile—dyehouses)	15–20	
Offices (public)	3	
Offices (private)	4	
Pickling plants	10–15	
Pump rooms	5	
Schools (grade)		15–25
Schools (high)		30–35
Restaurants	8–12	
Shops (machine)	5	
Shops (paint)	15–20	
Shops (railroad)	5	
Shops (woodworking)	5	
Substations (electric)	5–10	
Theaters		10–15
Turbine rooms (electric)	5–10	
Warehouses	2	
Waiting rooms (public)	4	

Data for Square-Type Louvers

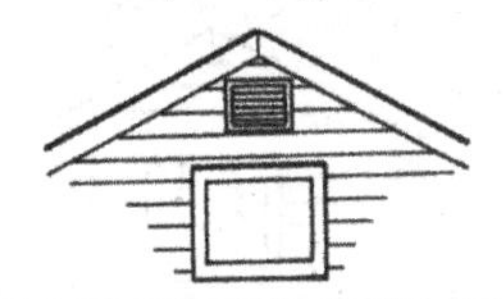

	Minimum Size of Square Outlet, (inches)		
	Metal Shutters		*Wood Slats*
Fan Diameter	*Automatic (90% Open Area)*	*Fixed (70% Open Area)*	*Fixed (60% Open Area)*
24″	26 × 26	32 × 32	34 × 34
30″	32 × 32	39 × 39	42 × 42
36″	38 × 38	45 × 45	49 × 49
42″	44 × 44	54 × 54	60 × 60
48″	50 × 50	62 × 62	68 × 68

(Courtesy Hayes-Albion Corporation)

Data for Triangular-Type Louvers

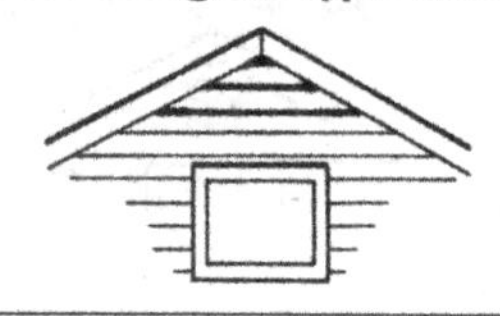

Fan Diameter	*Height of Triangular Louvers (for different roof pitches) 5/12 Pitch One Louver	6/12 Pitch One Louver	7/12 Pitch One Louver	8/12 Pitch One Louver
24"	2'0"	2'2"	2'4"	2'6"
30"	2'6"	2'9"	3'0"	3'3"
36"	3'0"	3'3"	3'6"	3'9"
42"	3'3"	3'9"	4'1"	4'4"
48"	3'10"	4'3"	4'7"	4'9"

*Heights given are for one triangular louvered opening only; when two openings are used, reduce heights by approximately 80%.
(Courtesy Hayes-Albion Corp.)

Ventilation Fan Terminology and Formulas

Air horsepower (ahp). The work done in moving a given volume (or weight) of air at a given speed. Air horsepower is also referred to as the Morse power output of a fan.

Area (A). The square feet of any plane surface or cross section.

Area of duct. The product of the height and width of the duct multiplied by the air velocity equals the cubic feet of air per minute flowing through the duct.

Brake horsepower (bhp). The work done by an electric motor in driving the fan, measured as horsepower delivered to the fan shaft. In belt-drive units, the total workload is equal to the workload of the electric motor plus the drive losses from belts and pulleys. The brake horsepower is always a higher number than air horsepower (ahp). Brake horsepower is also referred to as the *horsepower input* of the fan.

Cubic feet per minute (cfm). The physical volume of air moved by a fan per minute expressed as fan outlet conditions.

Density. The actual weight of air in pounds per cubic foot (0.075 at 70°F and 29.92 inches barometric pressure).

Fan inlet area. The inside area of the inlet collar.

Fan outlet area. The inside area of the fan outlet.

Mechanical efficiency (ME). A decimal number or a percentage representing the ratio of air horsepower (ahp) to brake horsepower (bhp) of a fan. It will always be less than 1.000 or 100 percent and may be expressed as follows:

$$\mathrm{ME} = \frac{\mathrm{ahp}}{\mathrm{bhp}}$$

Outlet velocity (OV). The outlet velocity of a fan measured in feet per minute.

Revolutions per minute (rpm). The speed at which a fan or motor turns.

Standard air. Air at 70°F and 29.92 inches barometric pressure weighing 0.075 pounds per cubic foot.

Static efficiency (SE). The static efficiency of a fan is the mechanical efficiency multiplied by the ratio of static pressure to the total pressure.

Static pressure (SP). The static pressure of a fan is the total pressure diminished by the fan velocity pressure. It is measured in inches of water. (See also *velocity pressure*).

Tip speed (TS). Also referred to as the peripheral velocity of wheel. It is determined by multiplying the circumference of the wheel by the rpm.

$$\text{TP} = \frac{\times \text{ wheel diameter in feet} \times \text{rpm}}{12}$$

$$\times \text{ wheel diameter in feet} \times \text{rpm}$$

The tip speed should not exceed 3300 rpm if quiet operation is desired.

Total pressure (TP). Any fan produces a total pressure (TP), which is the sum of the static pressure (SP) and the velocity pressure (VP). Total pressure represents the rise of pressure from fan inlet to fan outlet.

Velocity. The speed in feet per minute (fpm) at which air is moving at any location (for example, through a duct, inlet damper, outlet damper, or fan discharge point). When the performance data for air-handling equipment are given in feet per minute (fpm), conversion to cubic feet per minute (cfm) can be made by multiplying the fpm by the duct area:

$$\text{Air velocity} = 1000 \text{ fpm}$$

$$\text{Duct size} = 8 \text{ in.} \times 20 \text{ in.} = 160 \text{ in.}^2$$

$$\text{Duct area} = 160 \div 144 = 1.11 \text{ ft}^2$$

$$\text{Air flow} = 1000 \text{ fpm} \times 1.11 \text{ ft}^2 = 1110 \text{ cfm}$$

Velocity pressure (VP). Velocity pressure results only when air is in motion, and it is measured in inches of water. One-inch water gauge corresponds to 4005 fpm (standard air) velocity. The following formula is used for determining velocity pressure:

$$\text{VP} = \left[\frac{\text{air velocity}}{4005}\right]^2$$

Tips for Selecting Ventilation Fans

- A ½ horsepower, ⅓ horsepower, or ¼ horsepower 860-rpm, direct-drive fan used on three-phase motor voltages will usually eliminate single-phase magnetic hum.
- A belt-driven fan is less expensive, quieter, more flexible, and more adaptable to capacity change than a direct-drive fan.
- A propeller-type fan is recommended when operation offers little or no resistance, or when there is no duct system.
- A centrifugal or axial-flow fan should be used when a duct system is involved.
- Air should never be forced through ducts smaller than the area of the fan.
- A direct-drive fan motor usually has a long service life. The exception is the shaded-pole–type motor.

Minimum Fan Capacity (CFM) for Various Sections of the Country

Approxi. Volume of House (ft³)		*Minimum Fan Capacity Needed for Satisfactory Results, (CFM)*							
		North			*Central*			*South*	
3000		1000			2000		24″	3000	
4000		1320			2640			4000	
5000		1650			3300			5000	30″
6000		2000		24″	4000			6000	
7000		2310			4620	30″		7000	
8000		2540			5280		36′	8000	
9000		3000			6000			9000	
10,000		3330			6660			10,000	42″
11,000		3630			7260			11,000	
12,000	24″	4000		36″	8000			12,000	
13,000		4290			8580		48″	13,000	
14,000		4620	30″		9240			14,000	
15,000		5000			10,000	42″		15,000	
16,000		5280			10,560			16,000	
17,000		5610			11,220			17,000	
18,000		6000			12,000			18,000	
19,000		6270			12,540			19,000	
20,000		6660			13,320			20,000	
21,000		7000			14,000			21,000	
22,000		7260			14,520			22,000	

(Courtesy Hayes-Albion Corporation)

Recommended Dimensions for Attic Fan Exhaust Outlets

Recommended Dimensions of Attic Fan Exhaust Outlet		
Fan Diameter	***Air Delivery Range, (cfm)***	****Free Outlet Area Needed, (ft²)***
24″	3500/5000	4.70
30″	4500/8500	7.35
36″	8000/12000	10.06
42″	10000/15500	14.40
48″	12000/20000	18.85

**1.5 times fan area*
(Courtesy Hayes-Albion Corporation)

Rule-of-Thumb Sizing Method for Central Mechanical Ventilation

The fan in a central mechanical ventilation system should be sized to provide 0.047 cfm per square foot of house conditioned area. Always size the fan upwards in your calculations. The following table provides data for sizing central mechanical ventilation systems:

Area of Structure, ft²	*Total Ventilation, (cfm)*	*Minimum Duct Diameter, in.*
1300 or less	60	4
1301–1900	90	5
1901–2600	120	6
2601–3200	150	6
3201–3800	180	7
3801–4500	210	7
4501–5100	240	8

Sizing recommendations:

- Use the fan rating provided by the fan manufacturer.
- Calculate fan airflow (in cfm) to account for pressure loss due to ducting and fitting friction.

- Base system capacity on the actual installation.
- Size ducts according to the fan manufacturer's specifications.
- Use the fan manufacturer's recommendations for maximum duct length.
- Calculate the minimum main-trunk duct diameter on the basis of a maximum air velocity of 700 feet per minute.
- Size branch ducts based on the cfm serving each branch.

Calculating Fan Speed (rpm), Static Pressure (SP), and Brake Horsepower When Air Volume Moved by Fan Is Changed

1. Fan speed delivery will vary directly as the cfm ratio:

$$\text{New rpm} = \text{old rpm} \times \left[\frac{\text{new cfm}}{\text{old cfm}}\right]$$

2. Fan (and system) pressures will vary directly as the square of the rpm ratio:

$$\text{New SP (or TP or VP)} = \left[\frac{\text{new rpm}}{\text{old rpm}}\right]^2 \times \text{old SP(or TP or VP)}$$

3. Brake horsepower (bhp) on the fan motor (or air horsepower of the fan) will vary directly as the cube of the rpm ratio:

$$\text{New bhp (or ahp)} = \left[\frac{\text{new rpm}}{\text{old rpm}}\right]^3 \times \text{old bhp(or ahp)}$$

Example

A centrifugal fan delivers 10,000 cfm at a static pressure of 1.0 inch when operating at a speed of 600 rpm and requires an input of 3 hp. If 12,000 cfm is desired in the same installation, what will be the new fan speed (rpm), static pressure (SP), and horsepower (bhp) input? The three aforementioned formulas can be applied as follows:

$$1.\ \text{New rpm} = 600 \times \left[\frac{12{,}000}{10{,}000}\right]$$

$$= 600 \times 1.2 = 720$$

$$2.\ \text{New SP} = \left[\frac{720}{600}\right]^2 \times 1$$

$$= 1.44 \times 1 = 1.44$$

$$3.\ \text{New bhp} = \left[\frac{720}{600}\right]^3 \times 3$$

$$= 1.7 \times 3 = 5.1$$

Fan Calculation Formulas

1. A (area) × V (velocity) = cfm
2. cfm ÷ $V = A$
3. cfm ÷ $A = V$

Determining CFM by the Air-Change Method

In order to determine the required cfm for a structure of space by the air-change method, the following data are necessary:

1. The total cubic feet of air space in the structure or space.
2. The required number of air changes necessary to give satisfactory ventilation.

The total cubic feet of air space is easily determined by multiplying the dimensions of the structure of space. For example, a room 12 feet long and 10 feet wide with an 8-foot ceiling would have 960 cubic feet of air space (12 ft × 10 ft × 8 ft = 960 ft^3).

The required number of air changes necessary to give satisfactory ventilation will depend on a variety of factors, including (1) use, (2) number of people, (3) geographic location, and (4) height of ceiling.

Usually local health department codes will specify the required number of air changes for various installations.

Once the necessary data have been obtained, the following formula can be used to determine the cfm:

$$\text{cfm} = \frac{\text{building volume in cubic feet}}{\text{minutes air change}}$$

ELECTRIC MOTORS

Nominal Full-Load Ampere Ratings for Single-Phase Motors

		Full-Load Current	
HP	***RPM***	***115 V***	***230 V***
1⁄25	1550	1.0	0.5
1⁄25	1050	1.0	0.5
1⁄12	1725	2.0	1.0
	1140	2.4	1.2
	860	3.2	1.6
1⁄10	1550	2.4	1.2
1⁄8	1725	2.8	1.4
	1140	3.4	1.7
	860	4.0	2.0
1⁄6	1725	3.2	1.6
	1140	3.84	1.92
	860	4.5	2.25
1⁄4	1725	4.6	2.3
	1140	6.15	3.07
	860	7.5	3.75
1⁄3	1725	5.2	2.6
	1140	6.25	3.13
	860	7.35	3.67
1⁄2	1725	7.4	3.7
	1140	9.15	4.57
	860	12.8	6.4
3⁄4	1725	10.2	5.1
	1140	12.5	6.25
	860	15.1	7.55
1	1725	13.0	6.5
	1140	15.1	7.55
	860	15.9	7.95

(Courtesy Penn Ventilator Company, Inc.)

Nominal Full-Load Ampere Ratings for Three-Phase Motors

		Full-Load Current	
HP	**RPM**	***115 V***	***230 V***
	1725	0.95	0.48
¼	1140	1.4	0.7
	860	1.6	0.8
	1725	1.19	0.6
⅓	1140	1.59	0.8
	860	1.8	0.9
	1725	1.72	0.86
½	1140	2.15	1.08
	860	2.38	1.19
	1725	2.46	1.23
¾	1140	2.92	1.46
	860	3.26	1.63
	1725	3.19	1.6
1	1140	3.7	1.85
	860	4.12	2.06
	1725	4.61	2.31
1½	1140	5.18	2.59
	860	5.75	2.88
	1725	5.98	2.99
2	1140	6.50	3.25
	860	7.28	3.64
	1725	8.70	4.35
3	1140	9.25	4.62
	860	10.3	5.15
	1725	14.0	7.0
5	1140	14.6	7.3
	860	16.2	8.1
	1725	20.3	10.2
7½	1140	20.9	10.5
	860	23.0	11.5

(Courtesy Penn Ventilator Company, Inc.)

Circuit Wire Sizes for Individual Single-Phase Motors

Horsepower of Motor	*Volts*	*Approximate Starting Current Amperes*	*Approximate Full-Load Current Amperes*	*Length of Run in Feet (from Main Switch to Motor)*								
				Feet	*25*	*50*	*75*	*100*	*150*	*200*	*300*	*400*
¼	120	20	5	Wire Size	14	14	14	12	10	10	8	6
⅓	120	20	5.5	Wire Size	14	14	14	12	10	8	6	6
½	120	22	7	Wire Size	14	14	12	12	10	8	6	6
¾	120	28	21.5	Wire Size	14	12	12	10	8	6	4	4
¼	240	10	2.5	Wire Size	14	14	14	14	14	14	12	12
⅓	240	10	3	Wire Size	14	14	14	14	14	14	12	10
½	240	11	3.5	Wire Size	14	14	14	14	14	12	12	10
¾	240	14	4.7	Wire Size	14	14	14	14	14	12	10	10
1	240	16	5.5	Wire Size	14	14	14	14	14	12	10	10
1½	240	22	7.6	Wire Size	14	14	14	14	12	10	8	8
2	240	30	10	Wire Size	14	14	14	12	10	10	8	6
3	240	42	14	Wire Size	14	12	12	12	10	8	6	6
5	240	69	23	Wire Size	10	10	10	8	8	6	4	4
7½	240	100	34	Wire Size	8	8	8	8	6	4	2	2
10	240	130	43	Wire Size	6	6	6	6	4	4	2	1

Calculating Pulley Size and Compressor Speed for Belt-Driven Motors

The relative speeds of the motor and compressor are in direct relation to the size of the motor pulley and the compressor flywheel. The desired pulley size, or the resulting compressor speed, may be calculated from the following relation:

$$\text{Compressor rpm} = \frac{\text{motor-pulley diameter} \times \text{motor speed (rpm)}}{\text{compressor-pulley diameter}}$$

$$\text{Motor-pulley diameter} = \frac{\text{compressor speed (rpm)} \times \text{compressor-pulley diameter}}{\text{motor speed (rpm)}}$$

Example

Find the motor-pulley diameter required when a 1750-rpm motor is to be used to drive a compressor having an 8-inch diameter pulley when the compressor speed is 500 rpm.

Solution

The diameter of the motor pulley is obtained by substituting values as follows:

$$\text{Motor-pulley diameter} = \frac{500 \times 8}{1750} \text{ or } 2\tfrac{1}{4} \text{ in. (approximately)}$$

The desired driven speed in the preceding formula should be increased by 2 percent to allow for belt slip. The diameter of any suitable motor pulley at any motor speed may easily be calculated by inserting values of compressor-pulley diameter in a manner similar to the previous example.

Individual Branch Circuit Wiring for Single-Phase Induction Motors

		Copper Wire Size (minimum AWG no.)				
Motor Data		*Branch Circuit Length*				
Hp	*Volts*	*0–25 ft*	*50 ft*	*100 ft*	*150 ft*	*200 ft*
1/6	115	14	14	14	12	10
	230	14	14	14	14	14
1/4	115	14	14	12	10	8
	230	14	14	14	14	14
1/3	115	14	12	10	8	6
	230	14	14	14	14	12

(continued)

Motor Data		Copper Wire Size (minimum AWG no.)				
		Branch Circuit Length				
Hp	Volts	0–25 ft	50 ft	100 ft	150 ft	200 ft
½	115	14	12	10	8	6
	230	14	14	14	14	12
¾	115	12	10	8	6	4
	230	14	14	14	12	10
1	115	12	10	8	6	4
	230	14	14	14	12	10
1½	115	10	10	6	4	4
	230	14	14	12	10	8
2	115	10	8	6	4	
	230	14	12	12	10	8
3	115	6	6	4		
	230	10	10	10	8	8
5	230	8	8	8	6	4

Minimum Starting Torques for Squirrel-Cage Motors

No. of Poles	Percent of Full-Load Torque
2	150
4	150
6	135
8	125
10	120
12	115
14	110
16	105

Three-Phase Motor Weight and Horsepower

HP	Speed, (rpm)	Volts	NEMA Frame	Bearings	Therm. Prot.	F.L. Amps at 230 V	Est. Shpg. Wt. (lbs.)
¼	1725	230/460	48	Ball	None	1.5	19
	1140	230/460	56	Ball	None	1.3	22
	850	230/460	56	Ball	None	2.0	27
⅓	3450	230/460	48	Ball	None	1.7	19
	1725	230/460	56	Ball	None	1.5	22
	1140	230/460	56	Ball	None	1.7	25
½	3450	230/460	48	Ball	None	2.0	21
		230/460	56	Ball	None	2.0	21
	1725	230/460	56	Ball	None	2.2	22
	1425	220/380	56	Ball	None	2.3	20
	1140	230/460	56	Ball	None	2.6	27
¾	3450	230/460	56	Ball	None	2.6	24
	1725	200	56	Ball	None	3.6	24
		230/460	56	Ball	None	3.2	24
	1425	220/380	56	Ball	None	2.6	27
	1140	230/460	56	Ball	None	3.1	30
		230/460	143T	Ball	None	3.1	27

(Courtesy General Electric)

Formula for Calculating Speed of Induction Motors

The most common method used to calculate the speed of induction motors is by using the following formula:

$$\text{Slip}\ (\%) = \frac{(\text{synchronous speed} - \text{operating speed})\ 100}{\text{synchronous speed}}$$

The synchronous speed of a motor is found by the following:

$$N_S = \frac{\text{frequency} \times 120}{\text{number of poles}}$$

Example

A three-phase, squirrel-cage induction motor having four poles is operating on a 60-Hz AC circuit at a speed of 1728 rpm. What is the slip of this motor?

Solution

By substituting the values in the previous formulas,

$$N_S = \frac{60 \times 120}{4} = 1800 \text{ rpm}$$

$$\text{Slip} = \frac{(1800 - 1728)100}{1800} = 4\%$$

Basic Horsepower Formulas for Motor Applications

T = torque or twisting moment (force × moment arm length)

$\pi = 3.1416$

N = revolutions per minute

hp = horsepower (33,000 ft – lbs per min); applies to power output

R = radius of pulley, in feet

E = input voltage

I = current in amperes

P = power input in watts

$$\text{HP} = \frac{T(\text{lb} - \text{in.}) \times N(\text{rpm})}{63{,}025}$$

$$\text{HP} = T\,(\text{oz} - \text{in.}) \times N \times 9.917 \times 10^{-7}$$
$$= \text{approximately } T\,(\text{oz} - \text{in.}) \times N \times 10^{-6}$$

$$P = EI \times \text{power factor} = \frac{\text{hp} \times 746}{\text{motor efficiency}}$$

Formulas for Calculating Horsepower to Drive Pumps

$$\text{hp} = \frac{\text{gal per min} \times \text{total head (inc. friction)}}{3960 \times \text{eff. of pump}}$$

Where

Approximate friction head (ft)

$$= \frac{\text{pipe length (ft)} \times [\text{velocity of flow (fps)}]^2 \times 0.02}{5.367 \times \text{diameter (in.)}}$$

Eff. = Approx. 0.50 to 0.85

Formulas for Calculating Time to Change Speed of Rotating Mass

$$\text{Time (sec)} = \frac{WR^2 \times \text{change in rpm}}{308 \times \text{torque (ft} - \text{lb)}}$$

Where

$$WR^2 \text{ (disc)} = \frac{\text{weight (lb)} \times [\text{radius (ft)}]^2}{2}$$

WR^2 (rim)

$$= \frac{\text{wt. (lb)} \times [(\text{outer radius in ft})^2 + (\text{inner radius in ft})^2]}{2}$$

Formula for Calculating Horsepower to Drive Fans

$$\text{hp} = \frac{\text{cu ft air per min} \times \text{water gauge pressure (in.)}}{6.350 \times \text{eff.}}$$

Horsepower to kW Equivalents

Horsepower	Kilowatt (kW)*
1/20	0.025
	0.035
	0.05
	0.071
1/8	0.1
1/6	0.14
1/4	0.2
1/3	0.28
1/2	0.4
1	0.8
1½	1.1
2	1.6
3	2.5
5	4.0
7.5	5.6
10	8.0

*James W. Polk, *A Preview of Metric Motors*, Westinghouse Electric Corporation.

Horsepower/Watts vs. Torque Conversion Chart

		@ 1125 rpm		@ 1200 rpm		@ 1425 rpm	
Hp	**Watts**	**Oz.-in.**	**mN.in.**	**Oz.-in.**	**mN.m**	**Oz.-in.**	**mN.m**
1/2000	0.373	0.4482	3.1649	0.4202	2.9670	0.3538	2.4986
1/1500	0.497	0.5976	4.2198	0.5602	3.9561	0.4718	3.3314
1/1000	0.746	0.8964	6.3297	0.8403	5.9341	0.7077	4.9971
1/750	0.994	1.1951	8.4396	1.1205	7.9121	0.9435	6.6628
1/500	1.49	1.7927	12.6594	1.6807	11.8682	1.4153	9.9943
1/200	3.73	4.4818	31.6485	4.2017	29.6705	3.5383	24.9857
1/150	4.97	5.9757	42.1980	5.6023	39.5606	4.7177	33.3142
1/100	7.46	8.9636	63.2970	8.4034	59.3409	7.0765	49.9713
1/75	9.94	11.9515	84.3960	11.2045	79.1212	9.4354	66.6284
1/70	10.70	12.8052	90.4243	12.0048	84.7727	10.1093	71.3876
1/60	12.40	14.9393	105.4950	14.0056	98.9015	11.7942	83.2855
1/50	14.90	17.9272	126.5940	16.8086	118.6818	14.1531	99.9426
1/40	18.60	22.4090	158.2425	21.0085	148.3523	17.6913	124.9283
1/30	24.90	29.8787	210.9899	28.0113	197.8031	23.5884	166.5710
1/25	29.80	35.8544	253.1879	33.6135	237.3637	28.3061	199.8852
1/20	37.30	44.8180	316.4849	42.0169	296.7046	35.3827	249.8565
1/15	49.70	59.7574	421.9799	56.0225	395.6061	47.1769	333.1420
1/12	62.10	74.6967	527.4748	70.0282	494.5077	58.9711	416.4275
1/10	74.6	89.6361	632.9698	84.0338	593.4092	70.7653	499.7130
1/8	93.2	112.0451	791.2123	105.0423	741.7615	88.4566	624.6413
1/6	124.0	149.3934	1054.9497	140.0563	989.0153	117.9422	832.8550
1/4	186.0	224.0902	1582.4245	210.0845	1483.5230	176.9133	1249.2825
1/3	249.0	298.7869	2109.8994	280.1127	1978.0307	235.8844	1665.7101

		@ 1500 rpm		@ 1725 rpm		@ 1800 rpm	
Hp	**Watts**	**Oz.-in.**	**mN.m**	**Oz.-in.**	**mN.m**	**Oz.-in.**	**mN.m**
1/2000	0.373	0.3361	2.3736	0.2923	2.0640	0.2801	1.9780
1/1500	0.497	0.4482	3.1648	0.3897	2.7520	0.3735	2.6374
1/1000	0.746	0.6723	4.7473	0.5846	4.1281	0.5602	3.9561
1/750	0.994	0.8964	6.3297	0.7794	5.5041	0.7470	5.2747
1/500	1.490	1.3445	9.4945	1.1692	8.2561	1.1205	7.9121
1/200	3.730	3.3614	23.7364	2.9229	20.6403	2.8011	19.7803
1/150	4.97	4.4818	31.6485	3.8972	27.5204	3.7348	26.3737
1/100	7.46	6.7227	47.4727	5.8458	41.2806	5.6023	39.5606
1/75	9.94	8.9636	63.2970	7.7944	55.0409	7.4697	52.7475
1/70	10.70	9.6039	67.8182	8.3512	58.9723	8.0032	56.5152

(continued)

(continued)

		@ 1500 rpm		@ 1725 rpm		@ 1800 rpm	
Hp	*Watts*	*Oz.-in.*	*mN.m*	*Oz.-in.*	*mN.m*	*Oz.-in.*	*mN.m*
1/60	12.40	11.2045	79.1212	9.7431	68.8011	9.3371	65.9344
1/50	14.90	13.4454	94.9455	11.6917	82.5613	11.2045	79.1212
1/40	18.60	16.8068	118.6818	14.6146	103.2016	14.0056	98.9015
1/30	24.90	22.4090	158.2425	19.4861	137.6021	18.6742	131.8687
1/25	29.80	26.8908	185.8909	23.3833	165.1226	22.4090	158.2425
1/20	37.3	33.6135	237.3637	29.2292	206.4032	28.0113	197.8031
1/15	49.7	44.8180	316.4849	38.9722	275.2043	37.3484	263.7374
1/12	62.1	56.0225	395.6061	48.7153	344.0053	46.6854	329.6718
1/10	74.6	67.2270	474.7274	58.4583	412.8064	56.0225	395.6061
1/8	93.2	84.0338	593.4092	73.0729	516.0080	70.0282	494.5077
1/6	124.0	112.0451	791.2123	97.4305	688.0107	93.3709	659.3436
1/4	186.0	168.0676	1186.8184	146.1458	1032.0160	140.0563	989.0153
1/3	249.0	224.0902	1582.4245	194.8610	1376.0213	186.7418	1318.6871

		@ 3000 rpm		@ 3450 rpm		@ 3600 rpm	
Hp	*Watts*	*Oz.-in.*	*mN.m*	*Oz.-in.*	*mN.m*	*Oz.-in.*	*mN.m*
1/2000	0.373	0.1681	1.1868	0.1461	1.0320	0.1401	0.9890
1/1500	0.497	0.2241	1.5824	0.1949	1.3760	0.1867	1.3187
1/1000	0.746	0.3361	2.3736	0.2923	2.0640	0.2801	1.9780
1/750	0.994	0.4482	3.1648	0.3897	2.7520	0.3735	2.6374
1/500	1.490	0.6723	4.7473	0.5846	4.1281	0.5602	3.9561
1/200	3.730	1.6807	11.8682	1.4615	10.3202	1.4006	9.8902
1/150	4.97	2.2409	15.8242	1.9486	13.7602	1.8674	13.1869
1/100	7.46	3.3614	23.7364	2.9229	20.6403	2.8011	19.7803
1/75	9.94	4.4818	31.6485	3.8972	27.5204	3.7348	26.3737
1/70	10.70	4.8019	33.9091	4.1756	29.4862	4.0016	28.2576
1/60	12.40	5.6023	39.5606	4.8715	34.4005	4.6685	32.9672
1/50	14.90	6.7227	47.4727	5.8458	41.2806	5.6023	39.5606
1/40	18.6	8.4034	59.3409	7.3073	51.6008	7.0028	49.4508
1/30	24.9	11.2045	79.1212	9.7431	68.8011	9.3371	65.9344
1/25	29.8	13.4454	94.9455	11.6917	82.5613	11.2045	79.1212
1/20	37.3	16.8068	118.6818	14.6146	103.2016	14.0056	98.9015
1/15	49.7	22.4090	158.2425	19.4861	137.6021	18.6742	131.8687
1/12	62.1	28.0113	197.8031	24.3576	172.0027	23.3427	164.8359
1/10	74.6	33.6135	237.3637	29.2292	206.4032	28.0113	197.8031
1/8	93.2	42.0169	296.7046	36.5364	258.0040	35.0141	247.2538

(continued)

(continued)

Hp	Watts	@ 3000 rpm Oz.-in.	@ 3000 rpm mN.m	@ 3450 rpm Oz.-in.	@ 3450 rpm mN.m	@ 3600 rpm Oz.-in.	@ 3600 rpm mN.m
1/6	124.0	56.0225	395.6061	48.7153	344.0053	46.6854	329.6718
1/4	186.0	84.0338	593.4092	73.0729	516.0080	70.0282	494.5077
1/3	249.0	112.0451	791.2123	97.4305	688.0107	93.3709	659.3436

Hp	Watts	@ 5000 rpm Oz.-in.	@ 5000 rpm mN.m	@ 7500 rpm Oz.-in.	@ 7500 rpm mN.m	@ 10,000 rpm Oz.-in.	@ 10,000 rpm mN.m
1/2000	0.373	0.1008	0.7121	0.0672	0.4747	0.0504	0.3560
1/1500	0.497	0.1345	0.9495	0.0896	0.6330	0.0672	0.4747
1/1000	0.746	0.2017	1.4242	0.1345	0.9495	0.1008	0.7121
1/750	0.994	0.2689	1.8989	0.1793	1.2659	0.1345	0.9495
1/500	1.490	0.4034	2.8484	0.2689	1.8989	0.2017	1.4242
1/200	3.730	1.0084	7.1209	0.6723	4.7473	0.5042	3.5605
1/150	4.97	1.3445	9.4945	0.8964	6.3297	0.6723	4.7473
1/100	7.46	2.0168	14.2418	1.3445	9.4945	1.0084	7.1209
1/75	9.94	2.6891	18.9891	1.7927	12.6594	1.3445	9.4945
1/70	10.70	2.8812	20.3455	1.9208	13.5636	1.4406	10.1727
1/60	12.40	3.3614	23.7364	2.2409	15.8242	1.6807	11.8682
1/50	14.90	4.0336	28.4836	2.6891	18.9891	2.0168	14.2418
1/40	18.60	5.0420	35.6046	3.3614	23.7364	2.5210	17.8023
1/30	24.90	6.7227	47.4727	4.4818	31.6485	3.3614	23.7364
1/25	29.80	8.0672	56.9673	5.3782	37.9782	4.0336	28.4836
1/20	37.30	10.0841	71.2091	6.7227	47.4727	5.0420	35.6046
1/15	49.70	13.4454	94.9455	8.9636	63.2970	6.7227	47.4727
1/12	62.10	16.8068	118.6818	11.2045	79.1212	8.4034	59.3409
1/10	74.6	20.1681	142.4182	13.4454	94.9455	10.0841	71.2091
1/8	93.2	25.2101	178.0228	16.8068	118.6818	12.6051	89.0114
1/6	124.0	33.6135	237.3637	22.4090	158.2425	16.8068	118.6818
1/4	186.0	50.4203	356.0455	33.6135	237.3637	25.2101	178.0228
1/3	249.0	67.2270	474.7274	44.8180	316.4849	33.6135	237.3637

(Courtesy Bodine)

Formulas for Calculating Motor Horsepower from Meter Readings

Following are calculations to determine motor horsepower from meter readings.

DC Motors

$$\text{Hp} = \frac{\text{volts} \times \text{amperes} \times \text{efficiency}}{746}$$

Single-Phase AC Motors

$$\text{Hp} = \frac{\text{volts} \times \text{amperes} \times \text{efficiency} \times \text{power factor}}{746}$$

Two-Phase AC Motors

$$\text{Hp} = \frac{\text{volts} \times \text{amperes} \times \text{efficiency} \times \text{power factor} \times 2}{746}$$

Three-Phase AC Motors

$$\text{Hp} = \frac{\text{volts} \times \text{amperes} \times \text{efficiency} \times \text{power factor} \times 1.73}{746}$$

Formulas for Calculating Horsepower from Load (Mechanics' Data)

Following are calculations to determine horsepower from load.

Constant Speed

1. Rotational horsepower:

$$\text{Hp} = \frac{\text{torque (lb/ft)} \times \text{speed (rpm)}}{5250}$$

2. Prony-brake horsepower:

$$\text{Hp} = \frac{2 \times 3.1416 \times \text{lb applied at 1-ft radius} \times \text{rpm}}{33{,}000}$$

3. Linear horsepower:

$$\text{Hp} = \frac{\text{force (lb)} \times \text{velocity (ft/min)}}{33{,}000}$$

Acceleration from Zero to Full Speed

1. Rotational horsepower:

$$\text{Hp} = \frac{\text{inertia } (WR^2) \times \text{rpm}}{1.62 \times 106t \text{ (in seconds to come up to speed)}}$$

2. Linear horsepower:

$$\text{Hp} = \frac{\text{inertia } (W) \times \text{rpm}^2}{6.38 \times 107t \text{ (in seconds to come up to speed)}}$$

Formula for Calculating Horsepower for Compressing Gases

For one working stroke per revolution (single-acting):

$$\text{hp} = \frac{PLAN}{33{,}000 \times 0.90}$$

Where

P = effective pressure in cylinders, psi
L = length of stroke, ft
A = area of piston, in.^{2}
N = number of rpm

Conductivities (*k*), Conductances (*C*), and Resistances (*R*) of Various Building and Insulating Materials

Description					*Conductivity or Conductance*		*Resistance (R)*	
Material				*Density (lb per ft³)*	*(k)*	*(c)*	*Per Inch Thickness 1/k*	*For Thickness Listed 1/C*
Air Spaces	Position	Heat Flow	Thickness					
	Horizontal	Up (winter)	¾–4 in	—	—	1.18	—	0.85
	Horizontal	Up (summer)	¾–4 in	—	—	1.28	—	0.78
	Horizontal	Down (winter)	¾ in	—	—	0.98	—	1.02
	Horizontal	Down (winter)	1½ in	—	—	0.87	—	1.15
	Horizontal	Down (winter)	4 in	—	—	0.81	—	1.23
	Horizontal	Down (winter)	8 in	—	—	0.80	—	1.25
	Horizontal	Down (summer)	¾ in	—	—	1.18	—	0.85
	Horizontal	Down (summer)	1½ in	—	—	1.07	—	0.93
	Horizontal	Down (summer)	4 in	—	—	1.01	—	0.99
	Sloping, 45°	Up (winter)	¾–4 in	—	—	1.11	—	0.90
	Sloping, 45°	Down (summer)	¾–4 in	—	—	1.12	—	0.89
	Vertical	Horizontal (winter)	¾–4 in	—	—	1.03	—	0.97
	Vertical	Horizontal (summer)	¾–4 in	—	—	1.16	—	0.86

(continued)

(continued)

Material	Description		Density, (lb per ft³)	Conductivity or Conductance (k)	(c)	Resistance (R) Per Inch Thickness 1/k	For Thickness Listed 1/C
Air Surfaces	**Position**	**Heat Flow**					
Still air	Horizontal	Up	—	—	1.63	—	0.61
	Up sloping (45°)	Up	—	—	1.60	—	0.62
	Vertical	Horizontal	—	—	1.46	—	0.68
	Sloping (45°)	Down	—	—	1.32	—	0.76
	Horizontal	Down	—	—	1.08	—	0.92
15 mph wind	Any position—any direction (for winter)		—	—	6.00	—	0.17
7½ mph wind	Any position—any direction (for summer)		—	—	4.00	—	0.25
Building Board							
Boards, panels, sheathing, etc.							
	Gypsum or plaster board	⅜ in	50	—	3.10	—	0.32
	Gypsum or plaster board	½ in	50	—	2.25	—	0.45

	Plywood		34	0.80		1.25	—
	Plywood	¼ in	34	—	3.20	—	0.31
	Plywood	⅜ in	34	—	2.12	—	0.47
	Plywood	½ in	34	—	1.60	—	0.63
	Plywood or wood panels	¾ in	—	—	1.07	—	0.94
	Wood-fiber board,		26	0.42	—	2.38	—
	laminated or homogenous	31	0.50	—	2.00	—	
	Wood fiber—hardboard type		65	1.40	—	0.72	—
	Wood fiber—hardboard type	¼ in	65	—	5.60	—	0.18
	Wood—fir or pine sheathing	25/32 in	—	—	1.02	—	0.98
	Wood—fir or pine	1⅝ in	—	—	0.49	—	2.03
Building Paper	Vapor—permeable felt		—	—	16.70	—	0.06
	Vapor—seal, 2 layers of mopped 15-lb felt		—	—	8.35	—	0.12
	Vapor—seal, plastic film		—	—	—	—	Negl
Flooring Materials	Asphalt tile	⅛ in	120	—	24.80	—	0.04

(continued)

(continued)

Material	Description		Density, (lb per ft³)	Conductivity or Conductance (k)	(c)	Resistance (R) Per Inch Thickness 1/k	For Thickness Listed 1/C
	Carpet and fibrous pad		—	—	0.48	—	2.08
	Carpet and rubber pad		—	—	0.81	—	1.23
	Ceramic tile	1 in	—	—	12.50	—	0.08
	Cork tile		25	0.45		2.22	—
	Cork tile	⅛ in	—	—	3.60	—	0.28
	Felt, flooring		—	—	16.70	—	0.06
	Floor tile or linoleum—ave Value	⅛ in	—	20.00	0.05		
	Linoleum	⅛ in	80	—	12.00	—	0.08
	Plywood subfloor	⅝ in	—	—	1.28	—	0.78
	Rubber or plastic tile	⅛ in	110	—	42.40	—	0.02
	Terrazzo	1 in	—	—	12.50	—	0.98
	Wood, subfloor	25/32 in	—	—	1.02	—	0.98
	Wood, hardwood finish	¾ in	—	—	1.47	—	0.68
Insulating Materials	Cotton fiber		0.8–2.0	0.26	—	3.85	—

Blankets and batts	Mineral wool, fibrous form, processed from rock, slag, or glass		1.5–4.0	0.27	—	3.70	—
	Wood fiber		3.2–3.6	0.25	—	4.00	—
	Wood fiber, multilayer, stitched expanding		1.5–2.0	0.27	—	3.70	—
Board	Glass fiber		—	9.5	0.25	—	4.00
	Wood or cane fiber						
	Acoustical tile	½ in	—	0.84	—	1.19	—
	Acoustical tile	¾ in			0.56	—	1.78
	Interior finish (plank, tile, lath)		15.0	0.35	—	2.86	—
	Interior finish (plank, tile, lath)	½ in	15.0	—	0.70	—	1.43
	Roof deck slab						
	Approx.	1½ in	—	0.24	—	4.17	—
	Approx.	2 in	—	—	0.18	—	5.56
	Approx.	3 in	—	—	0.12	—	8.33
	Sheathing (impreg. or coated)		20.0	0.38	—	2.63	—
	Sheathing (impreg. or coated)	½ in	20.0	—	0.76	—	1.32
	Sheathing (impreg. or coated)	25/32 in	20.0	—	0.49	—	2.06

(continued)

(continued)

Description		Conductivity or Conductance			Resistance (R)	
Material		Density, (lb per ft³)	(k)	(c)	Per Inch Thickness 1/k	For Thickness Listed 1/C
Board and slabs	Cellular glass	9.0	0.40	—	2.50	—
	Corkboard (without added binder)	6.5–8.0	0.27	—	3.70	—
	Hog hair (with asphalt binder)	8.5	0.33	—	3.00	—
	Plastic (foamed)	1.62	0.29	—	3.45	—
	Wood, shredded (cemented in preformed slabs)	22.0	0.55	—	1.82	—
Loose fill	Macerated paper or pulp products	2.5–3.5	0.28	—	3.57	—
	Mineral wood (glass, slag, or rock)	2.0–5.0	0.30	—	3.33	—
	Sawdust or shavings	8.0–15.0	0.45	—	2.22	—
	Vermiculite (expanded)	7.0	0.48	—	2.08	—
	Wood fiber (redwood, hemlock, or fir)	2.0–3.5	0.30	—	3.33	—

Roof Insulation	All types						
	Preformed, for use above deck						
	Approx.	½ in			0.72	—	1.39
	Approx.	1 in			0.36	—	2.78
	Approx.	1½ in			0.24	—	4.17
	Approx.	2 in			0.19	—	5.26
	Approx.	2½ in			0.15	—	6.67
	Approx.	3 in			0.12	—	8.33
Masonry Materials	Cement mortar		116	5.0	—	0.20	—
Concretes							
	Gypsum-fiber concrete 87½% gypsum, 12½% wood chips		51	1.66	—	0.60	—
	Lightweight aggregates, including expanded		120	5.2	—	0.19	—
	shale, clay, or slate; expanded slags;		100	3.6	—	0.28	—
	cinders; pumice; perlite; vermiculite;		80	2.5	—	0.4	—
	also cellular concretes		60	1.7	—	0.59	
			40	1.15	—	0.86	
			30	0.90	—	1.11	

(continued)

(continued)

Material	Description		Density, (lb per ft^3)	Conductivity or Conductance (k)	Conductivity or Conductance (c)	Resistance (R) Per Inch Thickness 1/k	Resistance (R) For Thickness Listed 1/C
			20	0.70	—	1.43	
	Sand and gravel or stone aggregate (oven dried)		140	9.0	—	0.11	
	Sand and gravel or stone aggregate (not dried)		140	12.0	—	0.08	
	Stucco		116	5.0	—	0.20	
Masonry Units	Brick, common		120	5.0		0.20	
	Brick, face		130	9.0		0.11	
	Clay tile, hollow:						
	1 cell deep	3 in	—		1.25		0.80
	1 cell deep	4 in	—		0.90		1.11
	2 cells deep	6 in	—		0.66		1.52
	2 cells deep	8 in	—		0.54		1.85
	2 cells deep	10 in	—		0.45		2.22
	3 cells deep	12 in	—		0.40		2.50
	Concrete blocks, three oval core: sand and gravel aggregate						

		4 in	—	—	1.40	—	0.71
		8 in	—	—	0.90	—	1.11
		12 in	—	—	0.78	—	1.28
	Cinder aggregate	3 in	—	—	1.16	—	0.86
		4 in	—	—	0.90	—	1.11
		8 in	—	—	0.58	—	1.72
	Gypsum partition tile:						
	3 × 12 × 30-in solid						
			—	—	0.79	—	1.26
	3 × 12 × 30-in 4-cell		—	—	0.74	—	1.35
	3 × 12 × 30-in 3-cell						
			—	—	0.60	—	1.67
		3 in	—	—	0.79	—	1.27
	Lightweight aggregate	4 in	—	—	0.67	—	1.50
	(expanded shale, clay,						
	slate, or slag; pumice)						
		8 in	—	—	0.50	—	2.00
		12 in	—	—	0.44	—	2.27
	Stone, lime, or sand		—	12.50	—	0.08	—
Plastering	Cement plaster, sand		116	5.0	—	0.20	—
Materials	aggregate						
	Sand aggregate	½ in	—	—	10.00	—	0.10

(continued)

(continued)

Material	Description			Conductivity or Conductance		Resistance (R)	
			Density, (lb per ft^3)	(k)	(c)	Per Inch Thickness 1/k	For Thickness Listed 1/C
	Sand aggregate	¾ in	—	—	6.66	—	0.15
	Gypsum plaster						
	Lightweight aggregate	½ in	45	—	3.12	—	0.32
	Lightweight aggregate	⅝ in	45	—	2.67	—	0.39
	Lightweight aggregate on metal lath	¾ in	—	2.13	—	0.47	
	Perlite aggregate		45	1.5	—	0.67	—
	Sand aggregate		105	5.6	—	0.18	—
	Sand aggregate	½ in	105	—	11.10	—	0.09
	Sand aggregate	⅝ in	105	—	9.10	—	0.11
	Sand aggregate on metal lath	¾ in	—	—	7.70	—	0.13
	Sand aggregate on wood lath		—	—	2.50	—	0.40
	Vermiculate aggregate		45	1.7	—	0.59	—
Roofing	Asphalt roll roofing		70	—	6.50	—	0.15

	Asphalt shingles		70	—	2.27	—	0.44
	Built-up roofing	⅜ in	70	—	3.00	—	0.33
	Slate	½ in	—	—	20.00	—	0.05
	Sheet metal		—	400+	—	Negl	—
	Wood shingles		—	—	1.06	—	0.94
Siding Materials (On flat surface)	Shingles						
	Wood, 16-in 7½-in exposure		—	—	1.15	—	0.87
	Wood, double, 16-in, 12-in exposure		—	—	0.84	—	1.19
	Wood, plus insulation backer board	5/16 in	—	—	0.71	—	1.40
	Siding						
	Asphalt roll siding		—	—	6.50	—	0.15
	Asphalt insulating siding (½-in bd.)		—	—	0.69	—	1.45
	Wood, drop, 1 × 8 in		—	—	1.27	—	0.79
	Wood, bevel, ½ × 8 in, lapped		—	—	1.23	—	0.81

(continued)

(continued)

Material	Description		Conductivity or Conductance		Resistance (R)	
		Density, (lb per ft³)	(k)	(c)	Per Inch Thickness 1/k	For Thickness Listed 1/C
	Wood, bevel, ¾ × 10 in, lapped	—	—	0.95	—	1.05
	Wood, plywood, ⅜ in, lapped	—	—	1.59	—	0.59
	Structural glass	—	—	10.00	—	0.10
Woods	Maple, oak, and similar hardwoods	45	1.10	—	0.91	—
	Fir, pine, and similar softwoods	32	0.80	—	1.25	—

Courtesy ASHRAE 1960 Guide

Insulation Recommendations (R-Values) and Overall Coefficients of Heat Transfer (U-Values) for Major Areas of Heat Loss and Heat Gain in Residential Structures

Type of Construction	Opaque Sections Adjacent to Unheated Spaces			Opaque Sections Adjacent to Separately Heated Dwelling Units		
	Walls	***Floors***	***Ceilings***	***Walls***	***Floors***	***Ceilings***
Frame	*R*-11 *U*:0.07	*R*-11 *U*:0.07	*R*-19 *U*:0.05	*R*-11 *U*:0.07	*R*-11 *U*:0.07	*R*-11 *U*:0.07
Masonry	*R*-7 *U*:0.11	*R*-7 *U*:0.11	*R*-11 *U*:0.07	*R*-7 *U*:0.11	*R*-7 *U*:0.11	*R*-11 *U*:0.07
Metal section	*R*-11 *U*:0.07	Not applicable	Not applicable	*R*-11 *U*:0.07	Not applicable	Not applicable
Sandwich	*R*-[a] *U*:0.07	*R*-[a] *U*:0.07	*R*-[a] *U*:0.05	*R*-[a] *U*:0.07	*R*-[a] *U*:07	*R*-[a] *U*:07
Heated basement or unvented crawl space	*R*-7 *U*:0.11	Insulation not required	Insulation not required			
Unheated basement or vented crawl space	Insulation not required	See "Floors"	Insulation not required			

Insulation R-values refer to the resistance of the insulation only.

Courtesy Electric Energy Association.

[a]*Since the thermal resistance of sandwich construction depends upon its composition and thickness, the amount of additional insulation required to obtain the maximum U-factor must be calculated in each case.*

Coefficients of Heat Transmission for Solid Masonry Walls

Resistances Used		
Construction	**Resistance**	**(R)**
1. Outside surface (15-mph wind)		0.17
2. Face brick (4-in)		0.44
3. Common brick (4-in)		0.80
4. Air space		0.97
5. Gypsum lath (3/8-in)		0.32
6. Plaster (sand aggregate) (1/2-in)		0.09
7. Inside surface (still air)		0.68
Total resistance:		3.47
$U = 1/\mathbf{R} = 1/3.47 =$		0.29
See value 0.29 in boldface type in table below.		
Assume plain wall—no furring of or plaster.		
Total resistance:		3.47
Deduct 4. Air space	0.97	
5. Gypsum lath (3/8-in)	0.32	
6. Plaster (sand aggregate) (1/2-in)	0.09	1.38
Total resistance:		2.09
$U = 1/R = 1/2.09 =$		0.48

Exterior Construction		Interior Finish											No.
		None	Plaster 5/8-in on Plaster on Furring		Metal Lath and 3/4-in Plaster on Furring		Gypsum Lath (3/8-in) and 1/2-in Plaster on Furring				Insulation Board Lath (1/2-in) and 1/2-in Plaster on Furring		Wood Lath and 1/2-in Plaster (Sand agg.) 0.40	
Resistance			(Sand agg.) 0.11	(Lt. Wt. agg.) 0.39	(Sand agg.) 0.13	(Lt. Wt. agg.) 0.47	No plaster 0.32	(Sand agg.) 0.41	Lt. wt. agg.) 0.64	No plaster 1.43	(Sand agg.) 1.52		
Material	R	U	U	U	U	U	U	U	U	U	U	U	
		A	B	C	D	E	F	G	H	I	J	K	
Brick (face and common)													
(6 in)	0.61	0.68	0.64	0.54	0.39	0.34	0.36	0.35	0.33	0.26	0.25	0.35	1
(8 in)	1.24	0.48	0.45	0.41	0.31	0.28	0.30	0.29	0.27	0.22	0.22	0.29	2
(12 in)	2.04	0.35	0.33	0.30	0.25	0.23	0.24	0.23	0.22	0.19	0.19	0.23	3
(16 in)	2.84	0.27	0.26	0.25	0.21	0.19	0.20	0.20	0.19	0.16	0.16	0.20	4
Brick (common only)													
(8 in)	1.60	0.41	0.39	0.35	0.28	0.26	0.27	0.26	0.25	0.21	0.20	0.26	5
(12 in)	2.40	0.31	0.30	0.27	0.23	0.21	0.22	0.22	0.21	0.18	0.17	0.22	6
(16 in)	3.20	0.25	0.24	0.23	0.19	0.18	0.19	0.18	0.18	0.16	0.15	0.18	7

(continued)

(continued)

Exterior Construction		Interior Finish											No.
		None	Plaster 5/8-in on Plaster on Furring		Metal Lath and 3/4-in Plaster on Furring		Gypsum Lath (3/8-in) and 1/2-in Plaster on Furring				Insulation Board Lath (1/2-in) and 1/2-in Plaster on Furring		Wood Lath and 1/2-in Plaster (Sand agg.) 0.40	
Resistance			(Sand agg.) 0.11	(Lt. Wt. agg.) 0.39	(Sand agg.) 0.13	(Lt. Wt. agg.) 0.47	No plaster 0.32	(Sand agg.) 0.41	Lt. wt. agg.) 0.64	No plaster 1.43	(Sand agg.) 1.52		
Material	R	U	U	U	U	U	U	U	U	U	U	U	
		A	B	C	D	E	F	G	H	I	J	K	
Stone (lime and sand)													
(8 in)	0.64	0.67	0.63	0.53	0.39	0.34	0.36	0.35	0.32	0.26	0.25	0.35	8
(12 in)	0.96	0.55	0.52	0.45	0.34	0.31	0.32	0.31	0.29	0.24	0.23	0.31	9
(16 in)	1.28	0.47	0.45	0.40	0.31	0.28	0.29	0.28	0.27	0.22	0.22	0.29	10
(24 in)	1.92	0.36	0.35	0.32	0.26	0.24	0.25	0.24	0.23	0.19	0.19	0.24	11
Hollow clay tile													
(8 in)	1.85	0.36	0.36	0.32	0.26	0.24	0.25	0.25	0.23	0.20	0.19	0.25	12
(10 in)	2.22	0.33	0.31	0.29	0.24	0.22	0.23	0.22	0.21	0.18	0.18	0.23	13
(12 in)	2.50	0.30	0.29	0.27	0.22	0.21	0.22	0.21	0.20	0.17	0.17	0.21	14

Poured concrete													
30 lb/ft³													
(4 in)	4.44	0.19	0.19	0.18	0.16	0.15	0.15	0.15	0.14	0.13	0.13	0.15	15
(6 in)	6.66	0.13	0.13	0.13	0.12	0.11	0.11	0.11	0.11	0.10	0.10	0.11	16
(8 in)	8.88	0.10	0.10	0.10	0.09	0.09	0.09	0.09	0.09	0.08	0.08	0.09	17
(10 in)	11.10	0.08	0.08	0.08	0.08	0.07	0.08	0.08	0.07	0.07	0.07	0.08	18
80 lb/ft³													
(6 in)	2.40	0.31	0.30	0.27	0.23	0.21	0.22	0.22	0.21	0.18	0.17	0.22	19
(8 in)	3.20	0.25	0.24	0.23	0.19	0.18	0.19	0.18	0.18	0.16	0.15	0.18	20
(10 in)	4.00	0.21	0.20	0.19	0.17	0.16	0.16	0.16	0.15	0.14	0.14	0.16	21
(12 in)	4.80	0.18	0.17	0.17	0.15	0.14	0.14	0.14	0.14	0.12	0.12	0.14	22
140 lb/ft³													
(6 in)	0.48	0.75	0.69	0.58	0.41	0.36	0.38	0.37	0.34	0.27	0.26	0.37	23
(8 in)	0.64	0.67	0.68	0.53	0.39	0.34	0.36	0.35	0.32	0.26	0.25	0.35	24
(10 in)	0.80	0.61	0.57	0.49	0.36	0.32	0.34	0.33	0.31	0.25	0.24	0.33	25
(12 in)	0.96	0.55	0.52	0.45	0.34	0.31	0.32	0.31	0.29	0.24	0.23	0.31	26

Courtesy ASHRAE 1960 Guide

WEATHER DATA, DESIGN CONDITIONS, AND RELATED FACTORS USED IN HEATING AND COOLING LOAD CALCULATIONS

Atmospheric Pressure for Various Barometer Readings

Barometer (in Hg)	*Pressure (psi)*	*Barometer (in Hg)*	*Pressure (psi)*
28.00	13.75	29.75	14.61
28.25	13.88	29.921	14.696
28.50	14.00	30.00	14.74
28.75	14.12	30.25	14.86
29.00	14.24	30.50	14.98
29.25	14.37	30.75	15.10
29.50	14.49	31.00	15.23

Ground Temperatures Below the Frost Line

State	**City**	***Ground Temperature Commonly Used***
Alabama	Birmingham	66
Arizona	Tucson	60
Arkansas	Little Rock	65
California	San Francisco	62
	Los Angeles	67
Colorado	Denver	48
Connecticut	New Haven	52
District of Columbia	Washington	57
Florida	Jacksonville	70
	Key West	78
Georgia	Atlanta	65
Idaho	Boise	52
Illinois	Cairo	60
	Chicago	52
	Peoria	55
Indiana	Indianapolis	55
Iowa	Des Moines	52
Kentucky	Louisville	57
Louisiana	New Orleans	72
Maine	Portland	45
Maryland	Baltimore	57
Massachusetts	Boston	48
Michigan	Detroit	48
Minnesota	Duluth	41
	Minneapolis	44
Mississippi	Vicksburg	67
Missouri	Kansas City	57
Montana	Billings	42
Nebraska	Lincoln	52
Nevada	Reno	52
New Hampshire	Concord	47
New Jersey	Atlantic City	57
New Mexico	Albuquerque	57
New York	Albany	48
	New York City	52

State	City	Ground Temperature Commonly Used
North Carolina	Greensboro	62
North Dakota	Bismarck	42
Ohio	Cleveland	52
	Cincinnati	57
Oklahoma	Oklahoma City	62
Oregon	Portland	52
Pennsylvania	Pittsburgh	52
	Philadelphia	52
Rhode Island	Providence	52
South Carolina	Greenville	67
South Dakota	Huron	47
Tennessee	Knoxville	61
Texas	Abilene	62
	Dallas	67
	Corpus Christi	72
Utah	Salt Lake City	52
Vermont	Burlington	46
Virginia	Richmond	57
Washington	Seattle	52
West Virginia	Parkersburg	52
Wisconsin	Green Bay	44
	Madison	47
Wyoming	Cheyenne	42

Recommended Scale of Interior Effective Temperatures for Various Outside Dry-Bulb Conditions

Degrees Outside	*Degrees Inside*			
Dry-Bulb	***Dry-Bulb***	***Wet-Bulb***	***Dew Point***	***Effective Temperature***
100	82.5	69.0	62.3	76.0
95	81.0	67.7	60.8	74.8
90	79.5	66.5	59.5	73.6
85	78.1	65.3	58.0	72.5
80	76.7	64.0	56.6	71.3
75	75.3	63.0	55.6	70.2
70	74.0	62.0	54.5	69.0

Climatic Conditions for the United States[a,b]

Col. 1	Col. 2		Col. 3	Col. 4	Winter		Col. 5	Col. 6 Summer			Col. 7	Col. 8		
State and Station[c]	Latitude[d]		Elev.[e] ft	Median of Annual Extremes	99%	97½%	Coincident Wind Velocity[f]	Design Dry-Bulb			Outdoor Daily Range[g]	Design Wet-Bulb		
	Deg.	Min.						1%	2½%	5%		1%	2½%	5%
ALABAMA														
Birmingham AP	33	3	610	14	19	22	L	97	94	93	21	79	78	77
Huntsville AP	34	4	619	−6	13	17	L	97	95	94	23	78	77	76
Mobile CO	30	4	119	24	28	32	M	96	94	93	16	80	79	79
ALASKA														
Anchorage AP	61	1	90	−29	−25	−20	VL	73	70	67	15	63	61	59
Fairbanks AP	64	5	436	−59	−53	−50	VL	82	78	75	24	64	63	61
Kodiak	57	3	21	4	8	12	M	71	66	63	10	62	60	58
ARIZONA[b]														
Flagstaff AP	35	1	6973	−10	0	5	VL	84	82	80	31	61	60	59
Nogales	31	2	3800	15	20	24	VL	100	98	96	31	72	71	70
Yuma AP	32	4	199	32	37	40	VL	111	109	107	27	79	78	77

ARKANSAS														
Fayetteville AP	36	0	1253	3	9	13	M	97	95	93	23	77	76	75
Hot Springs Nat. Pk.	34	3	710	12	18	22	M	99	97	96	22	79	78	77
Texarkana AP	33	3	361	16	22	26	M	99	97	96	21	80	79	78
CALIFORNIA														
San Fernando	34	1	977	29	34	37	VL	100	97	94	38	73	72	71
San Francisco CO	37	5	52	38	42	44	VL	80	77	73	14	64	62	61
Yreka	41	4	2625	7	13	17	VL	96	94	91	38	68	66	65
COLORADO														
Alamosa AP	37	3	7536	–26	–17	–13	VL	84	82	79	35	62	61	60
Durango	37	1	6550	–10	0	4	M	88	86	83	30	64	63	62
Grand Junction AP	39	1	4849	–2	8	11	M	96	94	92	29	64	63	62
CONNECTICUT														
Hartford, Brainard Field	41	5	15	–4	1	5	M	90	88	85	22	77	76	74
New London	41	2	60	0	4	8	M	89	86	83	16	77	75	74
Windsor Locks, Bradley Field	42	0	169	–7	–2	2	M	90	88	85	22	76	75	73
DELAWARE														
Dover AFB	39	0	38	8	13	15	M	93	90	88	18	79	78	77
Wilmington AP	39	4	78	6	12	15	M	93	90	87	20	79	77	76

				Winter				Summer						
Col. 1	Col. 2		Col. 3	Col. 4			Col. 5	Col. 6			Col. 7	Col. 8		
State and Station[c]	Latitude[d]		Elev.[e] ft	Median of Annual Extremes	99%	97½%	Coincident Wind Velocity[f]	Design Dry-Bulb			Outdoor Daily Range[g]	Design Wet-Bulb		
	Deg.	Min.						1%	2½%	5%		1%	2½%	5%
DISTRICT OF COLUMBIA														
Andrews AFB	38	5	279	9	13	16	L	94	91	88	18	79	77	76
Washington National AP	38	5	14	12	16	19	VL	94	92	90	18	78	77	76
FLORIDA														
Key West AP	24	3	6	50	55	58	VL	90	89	88	9	80	79	79
Sarasota	27	2	30	31	35	39	M	93	91	90	17	80	80	79
Tallahassee AP	30	2	58	21	25	29	H	96	94	93	19	80	79	79
GEORGIA														
Americus	32	0	476	18	22	25	L	98	96	93	20	80	79	78
Dalton	34	5	702	10	15	19	L	97	95	92	22	78	77	76
Voldosta-Moody AFB	31	0	239	24	28	31	L	96	94	92	20	80	79	78
HAWAII														
Hilo AP	19	4	31	56	59	61	L	85	83	82	15	74	73	72
Honolulu AP	21	2	7	58	60	62	L	87	85	84	12	75	74	73
Wahiawa	21	3	215	57	59	61	L	86	84	83	14	75	74	73

IDAHO														
Burley	42	3	4180	−5	4	8	VL	95	93	89	35	68	66	64
Idaho Falls AP	43	3	4730r	−17	−12	−6	VL	91	88	85	38	65	64	62
Lewiston AP	46	2	1413	1	6	12	VL	98	96	93	32	67	66	65
ILLINOIS														
Carbondale	37	5	380	1	7	11	M	98	96	94	21	80	79	78
Freeport	42	2	780	−16	−10	−6	M	92	90	87	24	78	77	75
Peoria AP	40	4	652	−8	−2	2	M	94	92	89	22	78	77	76
INDIANA														
Jeffersonville	38	2	455	3	9	13	M	96	94	91	23	79	78	77
South Bend AP	41	4	773	−6	−2	3	M	92	89	87	22	77	76	74
Valparaiso	41	2	801	−12	−6	−2	M	92	90	87	22	78	76	75
IOWA														
Mason City AP	43	1	1194	−20	−13	−9	M	91	88	85	24	77	75	74
KANSAS														
Garden City AP	38	0	2882	−10	−1	3	M	100	98	96	28	74	73	72
Hutchinson AP	38	0	1524	−5	2	6	H	101	99	96	28	77	76	75
Wichita AP	37	4	1321	−1	5	9	H	102	99	96	23	77	76	75
KENTUCKY														
Ashland	38	3	551	1	6	10	L	94	92	89	22	77	76	75
Covington AP	39	0	869	−3	3	8	L	93	90	88	22	77	76	75
Hopkinsville, Campbell AFB	36	4	540	4	10	14	L	97	95	92	21	79	78	77

Col. 1	Col. 2		Col. 3	Col. 4			Col. 5	Col. 6			Col. 7	Col. 8		
				Winter				Summer						
State and Station[c]	Latitude[d]		Elev.[e] ft	Median of Annual Extremes	99%	97½%	Coincident Wind Velocity[f]	Design Dry-Bulb			Outdoor Daily Range[g]	Design Wet-Bulb		
	Deg.	Min.						1%	2½%	5%		1%	2½%	5%
LOUISIANA														
Alexandria AP	31	2	92	20	25	29	L	97	95	94	20	80	80	79
Natchitoches	31	5	120	17	22	26	L	99	97	96	20	81	80	79
New Orleans AP	30	0	3	29	32	35	M	93	91	90	16	81	80	79
MAINE														
Augusta AP	44	2	350	−13	−7	−3	M	88	86	83	22	74	73	71
Caribou AP	46	5	624	−24	−18	−14	L	85	81	78	21	72	70	68
Waterville	44	3	89	−15	−9	−5	M	88	86	82	22	74	73	71
MARYLAND														
Baltimore AP	39	1	146	8	12	15	M	94	91	89	21	79	78	77
Baltimore CO	39	2	14	12	16	20	M	94	92	89	17	79	78	77
Cumberland	39	4	945	0	5	9	L	94	92	89	22	76	75	74
MASSACHUSETTS														
Framingham	42	2	170	−7	−1	3	M	91	89	86	17	76	74	73
Greenfield	42	3	205	−12	−6	−2	M	89	87	84	23	75	74	73
New Bedford	41	4	70	3	9	13	H	86	84	81	19	75	73	72

MICHIGAN														
Detroit Met. CAP	42	2	633	0	4	8	M	92	88	85	20	76	75	74
Flint AP	43	0	766	−7	−1	3	M	89	87	84	25	76	75	74
Sault Ste. Marie AP	46	3	721	−18	−12	−8	L	83	81	78	23	73	71	69
MINNESOTA														
Alexandria AP	45	5	1421	−26	−19	−15	L	90	88	85	24	76	74	72
Bemidji AP	47	3	1392	−38	−32	−28	L	87	84	81	24	73	72	71
Minneapolis/St. Paul AP	44	5	822	−19	−14	−10	L	92	89	86	22	77	75	74
MISSISSIPPI														
Biloxi–Keesler AFB	30	2	25	26	30	32	M	93	92	90	16	82	81	80
Columbus AFB	33	4	224	13	18	22	L	97	95	93	22	79	79	78
Jackson AP	32	2	330	17	21	24	L	98	96	94	21	79	78	78
MISSOURI														
Kirksville AP	40	1	966	−13	−7	−3	M	96	94	91	24	79	78	77
Rolla	38	0	1202	−3	3	7	M	97	95	93	22	79	78	77
Sikeston	36	5	318	4	10	14	L	98	96	94	21	80	79	78
MONTANA														
Butte AP	46	0	5526r	−34	−24	−16	VL	86	83	80	35	60	59	57
Helena AP	46	4	3893	−27	−17	−13	L	90	87	84	32	65	63	61
Missoula AP	46	5	3200	−16	−7	−3	VL	92	89	86	36	65	63	61

Col. 1	Col. 2		Col. 3	Col. 4			Col. 5	Col. 6			Col. 7	Col. 8		
				Winter				Summer						
State and Station[c]	Latitude[d]		Elev.[e] ft	Median of Annual Extremes	99%	97½%	Coincident Wind Velocity[f]	Design Dry-Bulb			Outdoor Daily Range[g]	Design Wet-Bulb		
	Deg.	Min.						1%	2½%	5%		1%	2½%	5%
NEBRASKA														
Chadron AP	42	5	3300	−21	−13	−9	M	97	95	92	30	72	70	69
Hastings	40	4	1932	−11	−3	1	M	98	96	94	27	77	75	74
Kearney	40	4	2146	−14	−6	−2	M	97	95	92	28	76	75	74
NEVADA[b]														
Elko AP	40	5	5075	−21	−13	−7	VL	94	92	90	42	64	62	61
Las Vegas AP	36	1	2162	18	23	26	VL	108	106	104	30	72	71	70
Reno AP	39	3	4404	−2	2	7	VL	95	92	90	45	64	62	61
NEW HAMPSHIRE														
Berlin	44	3	1110	−25	−19	−15	L	87	85	82	22	73	71	70
Keene	43	0	490	−17	−12	−8	M	90	88	85	24	75	73	72
Portsmouth, Pease AFB	43	1	127	−8	−2	3	M	88	86	83	22	75	73	72
NEW JERSEY														
Atlantic City CO	39	3	11	10	14	18	H	91	88	85	18	78	77	76
Newark AP	40	4	11	6	11	15	M	94	91	88	20	77	76	75
Phillipsburg	40	4	180	1	6	10	L	93	91	88	21	77	76	75

NEW MEXICO														
Clovis AP	34	3	4279	2	14	17	L	99	97	95	28	70	69	68
Grants	35	1	6520	−15	−7	−3	VL	91	89	86	32	64	63	62
Las Cruces	32	2	3900	13	19	23	L	102	100	97	30	70	69	68
NEW YORK														
Geneva	42	5	590	−8	−2	2	M	91	89	86	22	75	73	72
Massena AP	45	0	202r	−22	−16	−12	M	86	84	81	20	75	74	72
NYC-Kennedy AP	40	4	16	12	17	21	H	91	87	84	16	77	76	75
NORTH CAROLINA														
Henderson	36	2	510	8	12	16	L	94	92	89	20	79	78	77
Rocky Mount	36	0	81	12	16	20	L	95	93	90	19	80	79	78
Wilmington AP	34	2	30	19	23	27	L	93	91	89	18	82	81	80
NORTH DAKOTA														
Bismarck AP	46	5	1647	−31	−24	−19	VL	95	91	88	27	74	72	70
Devil's Lake	48	1	1471	−30	−23	−19	M	93	89	86	25	73	71	69
Williston	48	1	1877	−28	−21	−17	M	94	90	87	25	71	69	67
OHIO														
Cincinnati CO	39	1	761	2	8	12	L	94	92	90	21	78	77	76
Norwalk	41	1	720	−7	−1	3	M	92	90	87	22	76	75	74
Toledo AP	41	4	646r	−5	1	5	M	92	90	87	25	77	75	74
OKLAHOMA														
Ardmore	34	2	880	9	15	19	H	103	101	99	23	79	78	77
Oklahoma City AP	35	2	1280	4	11	15	H	100	97	95	23	78	77	76
Woodward	36	3	1900	−3	4	8	H	103	101	98	26	76	74	73

				Winter				Summer						
Col. 1	Col. 2		Col. 3	Col. 4			Col. 5	Col. 6			Col. 7	Col. 8		
State and Station[c]	Latitude[d]		Elev.[e] ft	Median of Annual Extremes	99%	97 1/2%	Coincident Wind Velocity[f]	Design Dry-Bulb			Outdoor Daily Range[g]	Design Wet-Bulb		
	Deg.	Min.						1%	2 1/2%	5%		1%	2 1/2%	5%
OREGON														
Astoria AP	46	1	8	22	27	30	M	79	76	72	16	61	60	59
Baker AP	44	5	3368	−10	−3	1	VL	94	92	89	30	66	65	63
The Dalles	45	4	102	7	13	17	VL	93	91	88	28	70	68	67
PENNSYLVANIA														
Butler	40	4	1100	−8	−2	2	L	91	89	86	22	75	74	73
Sunbury	40	5	480	−2	3	7	L	91	89	86	22	76	75	74
West Chester	40	0	440	4	9	13	M	92	90	87	20	77	76	75
RHODE ISLAND														
Newport	41	3	20	1	5	11	H	86	84	81	16	75	74	73
Providence AP	41	4	55	0	6	10	M	89	86	83	19	76	75	74
SOUTH CAROLINA														
Charleston CO	32	5	9	23	26	30	L	95	93	90	13	81	80	79
Columbia AP	34	0	217	16	20	23	L	98	96	94	22	79	79	78
Rock Hill	35	0	470	13	17	21	L	97	95	92	20	78	77	76

SOUTH DAKOTA														
Aberdeen AP	45	3	1296	−29	−22	−18	L	95	92	89	27	77	75	74
Huron AP	44	3	1282	−24	−16	−12	L	97	93	90	28	77	75	74
Pierre AP	44	2	1718r	−21	−13	−9	M	98	96	93	29	76	74	73
TENNESSEE														
Bristol-Tri City AP	36	3	1519	−1	11	16	L	92	90	88	22	76	75	74
Chattanooga AP	35	0	670	11	15	19	L	97	94	92	22	78	78	77
Tullahoma	35	2	1075	7	13	17	L	96	94	92	22	79	78	77
TEXAS														
Brownsville AP	25	5	16	32	36	40	M	94	92	91	18	80	80	79
Killeen-Gray AFB	31	0	1021	17	22	26	M	100	99	97	22	78	77	76
Pampa	35	3	3230	0	7	11	M	100	98	95	26	73	72	71
UTAH														
Provo	40	1	4470	−6	2	6	L	96	93	91	32	67	66	65
St. George CO	37	1	2899	13	22	26	VL	104	102	99	33	71	70	69
Vernal AP	40	3	5280	−20	−10	−6	VL	90	88	84	32	64	63	62
VERMONT														
Barre	44	1	1120	−23	−17	−13	L	86	84	81	23	73	72	70
Burlington AP	44	3	331	−18	−12	−7	M	88	85	83	23	74	73	71
Rutland	43	3	620	−18	−12	−8	L	87	85	82	23	74	73	71

Col. 1	Col. 2		Col. 3	Col. 4	Winter		Col. 5	Summer Col. 6			Col. 7	Col. 8		
State and Station[c]	Latitude[d]		Elev.[e] ft	Median of Annual Extremes	99%	97½%	Coincident Wind Velocity[f]	Design Dry-Bulb			Outdoor Daily Range[g]	Design Wet-Bulb		
	Deg.	Min.						1%	2½%	5%		1%	2½%	5%
VIRGINIA														
Charlottesville	38	1	870	7	11	15	L	93	90	88	23	79	77	76
Harrisonburg	38	3	1340	0	5	9	L	92	90	87	23	78	77	76
Norfolk AP	36	5	26	18	20	23	M	94	91	89	18	79	78	78
WASHINGTON														
Bellingham AP	48	5	150	8	14	18	L	76	74	71	19	67	65	63
Moses Lake, Larson AFB	47	1	1183	−14	−7	−1	VL	96	93	90	32	68	66	65
Seattle CO	47	4	14	22	28	32	L	81	79	76	19	67	65	64
WEST VIRGINIA														
Elkins AP	38	5	1970	−4	1	5	L	87	84	82	22	74	73	72
Huntington CO	38	2	565r	4	10	14	L	95	93	91	22	77	76	75
Wheeling	40	1	659	0	5	9	L	91	89	86	21	76	75	74

WISCONSIN														
Ashland	46	3	650	−27	−21	−17	L	85	83	80	23	73	71	69
Beloit	42	3	780	−13	−7	−3	M	92	90	87	24	77	76	75
Sheboygan	43	4	648	−10	−4	0	M	89	87	84	20	76	74	72
WYOMING														
Casper AP	42	5	5319	−20	−11	−5	L	92	90	87	31	63	62	60
Cheyenne AP	41	1	6126	−15	−6	−2	M	89	86	83	30	63	62	61
Lander AP	42	5	5563	−26	−16	−12	VL	92	90	87	32	63	62	60

[a]Data for U.S. stations extracted from "Evaluated Weather Data for Cooling Equipment Design, Addendum No. 1, Winter and Summer Data," with the permission of the publisher, Fluor Products Company, Inc. Box 1267, Santa Rosa, CA.

[b]Data compiled from official weather stations, where hourly weather observations are made by trained observers and from other sources. Table prepared by ASHRAE Technical Committee 2.2, Weather Data and Design Conditions. Percentage of winter design data shows the percent of three-month period, December through February. Canadian data is based on January only. Percentage of summer design data shows the percent of a four-month period, June through September. Canadian data is based on July only.

[c]When airport temperature observations were used to develop design data, "AP" follows station name, and "AFB" follows Air Force Bases. Data for stations followed by "CD" came from office locations within an urban area and generally reflect an influence of the surrounding area. Stations without designation can be considered semi-rural and may be directly compared with most airport data.

[d]Latitude is given to the nearest 10 minutes, for use in calculating solar loads. For example, the latitude for Anniston, Alabama, is given as 33.4 or 33°40″.

[e]Elevations are ground elevations for each station as of 1964. Temperature readings are generally made at an elevation of 5 feet above ground, except for locations marked r, indicating roof exposure of thermometer.

[f]Coincident wind velocities derived from approximately coldest 600 hours out of 20,000 hours of December through February data per station. VL = Very Light, 70 percent or more of cold extreme hours ≤7 mph; M = Moderate, 50 to 74 percent cold extreme hours >7 mph; L = Light, 50 to 69 percent cold extreme hours ≤7 mph; H = High, 75 percent or more cold extreme hours >7 mph, and 50 percent are >12 mph.

[g]The difference between the average maximum and average minimum temperatures during the warmest month.

[h]More detailed data on Arizona, California, and Nevada may be found in "Recommended Design Temperatures, Northern California," published by the Golden Gate Chapter, and "Recommended Design Temperatures, Southern California, Arizona, Nevada," published by the Southern California Chapter.

Reduction in Standard Degree Days for All Areas Except the Pacific Coast

Degree Days at 65°F Base	Reduce by This Percentage	New Total
10,000	3.15	9685
9500	3.20	9196
9000	3.22	8710
8500	3.23	8226
8000	3.25	7740
7500	3.80	7215
7000	3.85	6720
6500	4.35	6218
6000	4.55	5727
5500	4.60	5247
5000	4.65	4768
4500	4.70	4289
4000	5.30	3788
3500	5.35	3313
3000	6.15	2816
2500	7.15	2321
2000	8.00	1840
1500	9.50	1358
1000	11.90	881
500	15.10	425

Courtesy National Oil Fuel Institute

Reduction in Degree-Day (DD) Base When Calculated Heat Loss (Btu per Degree Temperature Difference) is Less Than 1000

Calculated Heat Loss	*Revised DD Base, °F*
200	55
300	60
400	61
500	62
600	63
700	64
800	64
900	64
1000	65

Courtesy National Oil Fuel Institute

Shade Factors for Various Types of Shading

Type of Shading	*Finish on Side Exposed to Sun*	*Shade Factor*
Canvas awning sides open	Dark or medium	0.25
Canvas awning top and sides tight against building	Dark or medium	0.35
Inside roller shade, fully drawn	White, cream	0.41
Inside roller shade, fully drawn	Medium	0.62
Inside roller shade, fully drawn	Dark	0.81
Inside roller shade, half drawn	White, cream	0.71
Inside roller shade, half drawn	Medium	0.81
Inside roller shade, half drawn	Dark	0.91
Inside venetian blind, slats set at 45°	White, cream	0.56
Inside venetian blind, slats set at 45°	Diffuse reflecting aluminum metal	0.45
Inside venetian blind, slats at 45°	Medium	0.65
Inside venetian blind, slats set at 45°	Dark	0.75
Outside venetian blind, slats set at 45°	White, cream	0.15
Outside venetian blind, slats set at 45° extended as awning fully covering window	White, cream	0.15
Outside venetian blind, slats set at 45° extended as awning covering ⅔ of window	White, cream	0.43

	Dark	*Green Tint*
Outside shading screen, solar altitude 10°	0.52	0.46
Outside shading screen, solar altitude 20°	0.40	0.35
Outside shading screen, solar altitude 30°	0.25	0.24
Outside shading screen, solar altitude, above 40°	0.15	0.22

(Courtesy ASHRAE 1960 Guide)

Heat Gain Due to Solar Radiation (Single Sheet of Unshaded Common Window Glass)

	Sun Time		Instantaneous Heat Gain in Btu per Hour (ft^2)								
Latitude	A.M. →	P.M. ↓	N	NE	E	SE	S	SW	W	NW	Horiz.
30° north	6 A.M.	6 P.M.	25	98	108	52	5	5	5	5	17
	7	5	23	155	190	110	10	10	10	10	71
	8	4	16	148	205	136	14	13	13	13	137
	9	3	16	106	180	136	21	15	15	15	195
	10	2	17	54	128	116	34	17	16	16	241
	11	1	18	20	59	78	45	19	18	18	267
	12		18	19	19	35	49	35	19	19	276
40° north	5 A.M.	7 P.M.	3	7	6	2	0	0	0	0	1
	6	6	26	116	131	67	7	6	6	6	25
	7	5	16	149	195	124	11	10	10	10	77
	8	4	14	129	205	156	18	12	12	12	137
	9	3	15	79	180	162	42	14	14	14	188
	10	2	16	31	127	148	69	16	16	16	229
	11	1	17	18	58	113	90	23	17	17	252
	12		17	17	19	64	98	64	19	17	259

Latitude	Sun Time A.M. →	Sun Time P.M. ↓	N	NE	E	SE	S	SW	W	NW	Horiz.
			Instantaneous Heat Gain in Btu per Hour (ft^2)								
50° north	5 A.M.	7 P.M.	20	54	54	20	3	3	3	3	6
	6	6	25	128	148	81	8	7	7	7	34
	7	5	12	139	197	136	12	10	10	10	80
	8	4	13	107	202	171	32	12	12	12	129
	9	3	14	54	176	183	72	14	14	14	173
	10	2	15	18	124	174	110	16	15	15	206
	11	1	16	16	57	143	136	42	16	16	227
	12		16	16	18	96	144	96	18	16	234
		↑ P.M. →	N	NE	E	SE	S	SW	W	NW	Horiz.

(Courtesy ASHRAE 1960 Guide)

Atmospheric Pressure and Barometer Readings for Various Altitudes

Altitude above Sea Level, (Feet)	*Atmospheric Pressure (Pounds per Square Inch)*	*Barometer Reading (Inches of Mercury)*
0	14.69	29.92
500	14.42	29.38
1000	14.16	28.86
1500	13.91	28.33
2000	13.66	27.82
2500	13.41	27.31
3000	13.16	26.81
3500	12.92	26.32
4000	12.68	25.84
4500	12.45	25.36
5000	12.22	24.89
5500	11.99	24.43
6000	11.77	23.98
6500	11.55	23.53
7000	11.33	23.09
7500	11.12	22.65
8000	10.91	22.22
8500	10.70	21.80
9000	10.50	21.38
9500	10.30	20.98
10,000	10.10	20.58
10,500	9.90	20.18
11,000	9.71	19.75
11,500	9.52	19.40
12,000	9.34	19.03
12,500	9.15	18.65
13,000	8.97	18.29
13,500	8.80	17.93
14,000	8.62	17.57
14,500	8.45	17.22
15,000	8.28	16.88

Infiltration Rate Through Various Types of Windows*

Type of Window	Remarks	Wind Velocity, mph					
		5	10	15	20	25	30
Double-Hung Wood sash Windows (Unlocked)	Around frame in masonry wall—not caulked	3	8	14	20	27	35
	Around frame in masonry wall—caulked	1	2	3	4	5	6
	Around frame in wood-frame construction	2	63	11	17	23	30
	Total for average window, non-weather-stripped, $1/16$-in. crack and $3/64$-in. clearance; includes wood frame leakage	7	21	39	59	80	104
	Ditto, weather-stripped	4	13	24	36	49	63
	Total for poorly fitted window, non-weather-stripped, $3/32$-in. crack and $3/32$-in. clearance; includes wood frame leakage	27	69	111	154	199	249
	Ditto, weather-stripped	6	19	34	51	71	92
Double-Hung Metal Windows	Non-weather-stripped, locked	20	45	70	96	125	154
	Non-weather-stripped, unlocked	20	47	74	104	137	170
	Weather-stripped, unlocked	6	19	32	46	60	76

Rolled Section Steel Sash	Industrial pivoted, 1/16-in. crack	52	108	176	244	304	372
	Architectural projected, 1/32-in. crack	15	36	62	86	112	139
	Architectural projected, 3/64-in. crack	20	52	88	116	152	182
	Residential casement, 1/64-in. crack	6	18	33	47	60	74
	Residential casement, 1/32-in. crack	14	32	52	76	100	128
	Heavy casement section, projected, 1/64-in. crack	3	10	18	26	36	48
	Heavy casement section, projected, 1/32-in. crack	8	24	38	54	72	92
Hollow metal, Vertically pivoted window		30	88	145	186	221	242

**Expressed in cubic feet per foot of crack per hour. The infiltration rate through cracks around closed doors is generally estimated at twice that calculated for a window.*

Courtesy ASHRAE 1960 Guide

Infiltration Through Various Types of Wall Construction

Type of Wall	*Wind Velocity, mph*					
	5	10	15	20	25	30
Brick wall						
8½-in. plain	2	4	8	12	19	23
Plastered	0.02	0.04	0.07	0.11	0.16	0.24
Plain	1	4	7	12	16	21
13-in. plastered	0.01	0.01	0.03	0.04	0.07	0.10
Frame wall, lath and plaster	0.03	0.07	0.13	0.18	0.23	0.26

Courtesy ASHRAE 1960 Guide